Agriculture : Bio-Integrated Farming Systems

AGRICULTURE
Bio-Integrated Farming Systems

K.P. Roy

AMIGA PRESS. INC
Delhi -110051

AMIGA
PRESS INC.

AMIGA PRESS INC.
Registered Office:
19 DBS Building, Ansari Road, Darya Ganj, Delhi. 110002, India
Phone: +91-11-23283333 Email: amigapress@gmail.com

This edition is published by Amiga Press.Inc

ISBN : 978-93-87295-00-1

First Published 2018
Reprint 2026

Printed and bound in India by Replika Press Pvt. Ltd.

Preface

The Bio-Integrated Farm is a must-read, twenty-first-century manual for managing these natural resources and brings system farming and permaculture to a whole new level.

Farming enterprises include crop, livestock, poultry, fish, tree crops, plantation crops, etc. A combination of one or more enterprises with cropping, when carefully chosen, planned and executed, gives greater dividends than a single enterprise, especially for small and marginal farmers. Farm as a unit is to be considered and planned for effective integration of the enterprises to be combined with crop production activity.

In many parts of Asia, the productive use of land and water resources has been integrated by transforming wetlands into ponds separated by cultivable ridges. Overall integrated farming systems that include semi-intensive aquaculture are less risky for the resource-poor farmer than intensive fish farms, because of their efficiency derived from synergism, their diversity of produce, and their environmental soundness. In many traditional systems aquaculture goes beyond fish production and cash income as pond water and pond biota perform many ecological, social, and cultural services on an integrated farm. Thus aquaculture and water management act as an engine driving the sustainability of the entire farming system.

Bio-pesticides have been trialed and classified to increase production. The following basic raw material have been collected and carefully processed for use in the demonstration farm: Wettable powder (dried pesticide for convenience of storage and transportation); Liquid based formulations 20L (broad spectrum) and Moses's own "Trust Me" pesticide which contains lantana camara, neem, pawpaw leaves, nerium, chilli, cow urine, cow dung, garlic and jatropha curcas.

Traditional integrated biosystems often use labour-intensive bio-systems or technologies in low-input, organic farming that can provide a variety of products at a micro-level. They recycle nutrients in solids and liquids into biofertilisers, feeds, aquatic plants, fish and food. In the past century, access to chemical fertilisers, pesticides, herbicides and mechanization led to monoculture crop production. This replaced the practice of the traditional integrated farming systems in many countries. In the new millennium, we will continue to face challenges at a larger dimension in scales. They include : local shortages and/or need for more efficient use of natural resources (water, forest, energy, agricultural land), managing biodegradable wastes from cities and farms in an environmentally sound manner, ensuring food security, reducing poverty and creating sustainable livelihoods and human communities.

In this book an effort has been made to present the detailed information about the integrated farming system concept principles components, model development and their usefulness to the farm household. We hope that the book would prove to be useful to the students, researchers, planners, farmers and other stakeholders.

—*Editor*

Contents

1

Agriculture in India

India has made lot of progress in agriculture since independence in terms of growth in output, yields and area under many crops. It has gone through a green revolution, a white revolution, a yellow revolution and a blue revolution. Today, India is the largest producer of milk, fruits, cashewnuts, coconuts and tea in the world, the second largest producer of wheat, vegetables, sugar and fish and the third largest producer of tobacco and rice. The per capita availability of foodgrains has risen in the country from 350 gm in 1951 to about 500 gm per day now, of milk from less than 125 gm to 210 gm per day and of eggs from 5 to 30 per annum despite the increase in population from 35 crores to 95 crores. However, India would have been in an even better position now both in terms of agricultural output and economic development had our planners given the required importance to its development it deserved in the early years since independence.

Even today, the farmers in India are able to obtain only 15 per cent of their requirements of agricultural credit from banks. The various state seed corporations are able to produce only 1 0 per cent of the seeds required by our farmers. At present only 2330 per cent of the farmers are able to derive any benefits of extension services provided by various government agencies and every year about 20 per cent of the crop is lost due to mishandling, spillage, floods, droughts and pests and diseases.

In fruits and vegetables the loss is around 30 per cent. The farmers are still being exploited by moneylenders who provide finance at exorbitant interest rates and there are cartels of traders who pay very little for their produce even in the well recognized mandies in the country.

AGRICULTURE – SUB-SECTORS

Indian agriculture broadly consists of four sub-sectors. Agriculture proper including all food-crops oilseeds, fiber, plantation crops, fruits and vegetables is the largest accounting for nearly 70 percent of the agriculture sector as a whole. The rapid growth in this sub-sector through exploitation of wastelands and fallows, spread of irrigation and adoption of production enhancing technologies was critical in transforming India from a country vulnerable to food shortages to one of exportable surplus. Although this sub-sector has made impressive progress its share in the sector as a whole has declined from 78 percent in 1960-61 to less than 70 percent by early 90s.

Table. Agriculture sub-sectors – value of output (%)

Sub-sector	1960-61	1970-71	1980-81	1993-94	2014-15
Agriculture	78.0	79.3	75.3	68.5	70.0
Livestock	16.6	15.1	17.2	24.5	25.0
Forestry	4.4	4.4	5.9	3.8	3.0
Fisheries	1.0	1.2	1.6	3.2	2.0
	100.0	100.0	100.0	100.0	100.0
Total value at Current price (Rs. Billion)	89	218	615	2740	2740

Correspondingly the share of livestock sector has increased considerably. The livestock industry has grown from Rs. 15 billion in early 1960s to Rs. 100 billion by 1980-81 and Rs. 672 billion by 1993-94. In nominal terms the sector grew at almost 15 percent per annum during 1980s. Milk production, which was almost stagnant for two decades ending 1970, grew by over 5 percent per annum in the 80s. Similarly, production of eggs increased at the rate of about 6.5 percent during the same period. As a result the share of livestock increased from about 17 percent till early 80s to 25 percent by 1993-94.

Though it plays relatively a minor role within the sector as a whole, fishing sub-sector activities have been on the rise. The sub-sector has grown from only Rs. 3 billion in 1970-71 to nearly Rs. 90 billion in 1993-94. The growth was particularly rapid in 70s and 80s. Value added increased at over 5 percent per annum during this period. In real terms forestry and logging activities have been on the decline since mid seventies. As of 1993-94, the size of the industry in terms of value of output was 103 billion.

Over the past three decades, the country has successfully transformed itself from a food deficit economy to one which is essentially self sufficient in availability of foodgrains and other essential commodities, albeit only at the prevailing level of effective demand. Annual aggregate foodgrains production, which averaged about 82 million tonnes in 1960-61 increased to 123.7 and 172.5 million tonnes for the trienniums ending 1980-81 and 1990-91 respectively.

Current (1998-99) production level is 195 million tonnes and the country has been able to accumulate substantial, (35 million tonnes) stocks of foodgrains to cope up with any sudden difficulties arising from drought or a similar situation in any part of the country.

Table. Foodgrains Production

Year	Area (million ha)	Production (million tonnes)	Yield kg/ha
1950-51	97.32	50.82	522
1960-61	115.58	82.02	709
1970-71	124.32	108.42	872
1980-81	126.67	129.59	1023
1990-91	127.84	176.39	1379
1998-99	124.0	195.25	1574
2014-15	124.0	197.25	1600

Increased outputs, have been achieved chiefly by adopting, since mid sixties, a strategy aimed at increasing foodgrains production by concentrating public sector efforts and resources in regions with a high potential for quick and substantial productivity gains through increased cropping intensity and average yields. These were the areas favoured by agroclimatic resource conditions and where irrigation facilities already existed or could be developed relatively rapidly. The main elements of this strategy were:

(i) expansion of irrigation coverage,

(ii) increased provision and utilization of key inputs – mainly high yielding varieties (HYVs) of crops, mainly of wheat and rice and chemical fertilizers and plant protection chemicals,

(iii) expansion and improvement of institutional support services such as research and extension and

(iv) price policies favourable to producers of major foodgrains.

The success of this strategy was made possible by development and availability of replicable production technology packages, so called 'Green Revolution' technologies. Irrigation facilitated double cropping and widespread adoption of HYVs. The HYVs performed particularly well under irrigated conditions, were highly responsive to fertilizers and their short duration permitted increases in cropping intensities.

Irrigation development was the cornerstone of the strategy. Undivided India was amongst the largest irrigated areas in the world. With partition nearly one-third of the irrigated area went to Pakistan. At the time of independence the net irrigated area was 20.9 million ha (gross irrigated area 22.6 million ha). Recognizing large-scale development of irrigation facilities as critical to rapid agricultural growth, the country has spent about Rs. 45,000 crores on irrigation development in the first four decades after independence. During the period 1950-51 to 1965-66 development of irrigation through government canals grew from 7.2 million ha to 9.8 million ha – a growth rate of 2.1 percent per annum. During 1970s this pace dropped slightly to 1.9 percent. In 1980s the rate of increase dropped significantly to 1.1 percent per annum.

The growth of tube-well irrigation, however, increased rapidly from 4.5 million ha in 1970-71 to 9.5 million ha in 1980-81 and then to 14.3 million ha by 1990-91. The net irrigated area increased from 31 million ha in 1970-71 to 53.5 million ha in 1995-96 which corresponds to 22 percent of the net sown area in 1970-71 and 37.63 percent in 1995-96 *(Tables 3, 4).* With improvements in irrigation efficiency the gross irrigated areas has increased to 71.51 million ha. The percentage of gross cropped area service by irrigation increased from 18.3 percent in 1960-61 to 23.0 percent in 1970-71 and to over 38 percent at present.

Table. India – irrigated area

Year	Net	Gross	Net as % of net sown	Gross as % of gross cropped
	——— million ha ———		———— % ————	
1970-71	31.10	38.19	22.17	23.04
1980-81	38.72	49.78	27.66	28.83
1990-91	48.02	63.20	33.61	34.03
1995-96	53.51	71.51	37.63	38.33
2010-11	56.00	72.00	38.00	39.00

Table. Sources of irrigation

Year	Net Irrigated Area (million ha)	Sources				
		Canals	Tanks	Tube-wells	Other wells	Other Sources
		(% net irrigated area)				
1960-61	24.80	10.4	4.6	0.2	7.2	2.4
1970-71	31.10	41.28	13.22	14.34	23.88	7.29
1980-81	38.72	39.49	8.22	24.62	21.08	6.59
1990-91	48.02	36.34	6.13	29.69	21.73	6.11
1995-96	53.01	32.04	5.81	33.52	22.16	6.47
2010-11	55.00	33.00	6.00	34.00	23.00	6.66

Fertilizers have constituted yet another key input in addition to expanded irrigation and spread of HYVs in achieving goals of high production and productivity. India currently occupies third position in the world, after China and USA, in terms of fertilizer production and consumption. Consumption of fertilizers has increased from 1.54 million tonnes in 1967-68, representing thepre green revolution era, to 17.31 million tonnes currently (1997-98).

The average per hectare use of fertilizers currently around 85 kg per hectares is the lowest among several Asian countries. However, rice and wheat account for a major fraction, around 65 percent of the total fertilizer consumed in the country, with very little fertilizers going to the rainfed areas. According to some current projections, fertilizer's use will need to increase to 30-35 million tonnes to meet the foodgrains need of 2020. The demand for nutrients will stretch by almost another 15 million tonnes if requirements for horticulture, vegetables and plantation and commercial crops are included. At present domestic production of N and P fertilizers (13.42 million tonnes) falls short of consumption by over 20 percent. In addition the entire requirement of K fertilizer is imported. Most agricultural development programmes initiated in 1960s were concentrated in regions of high potential. Thus five states, Punjab, Haryana, Uttar Pradesh, Andhra Pradesh and Tamil Nadu account for 50 percent of the country's net irrigated and 53 percent of the gross irrigated area. The combination of expanding irrigation coverage and widespread adoption of short duration HYVs led to significant increases in cropping intensities. Acreage cropped more than once per year increased from 13 million ha in 1950-51 to about 44 million ha at present. Average cropping intensity for the country as a whole rose from 115 percent in 1960-61 to 131 in 1993-94. By 1993-94 cropping intensity has risen to 187 percent in Punjab, 167 percent in Haryana and 142 percent in Uttar Pradesh.

Table. Cropping intensity

Years	Net sown area	Gross cropped area	Area cropped more than once	Cropping intensity
	million ha			(%)
1950-51	118.8	131.9	13.1	111.1
1960-61	133.2	152.8	19.6	114.3
1970-71	140.3	165.8	25.5	118.2
1980-81	140.0	172.6	32.6	123.3
1990-91	143.0	185.7	42.7	129.9
1993-94	142.0	186.4	44.4	131.2
2010-11	165.0	190.0	45.0	132.6

An important consequence of the strategies adopted since sixties has been to boost production of, chiefly, two crops rice and wheat. Their share in total foodgrains production went up from 57 percent in 1970-71 to more than 75 percent in 1990-91. Production of foodgrains other than rice and wheat did not increase significantly and in the eastern region even the yield of rice did not increase. Agricultural production and income rose substantially in the north-western states of Punjab, Haryana, Western Uttar Pradesh, parts of Rajesthan, Tamil Nadu and Andhra Pradesh. By contrast productivity and output growth have been modest in eastern and central India and in deccan plateau. Progress was particularly slow in rainfed areas, which account for over 60 percent of the cropped area and where a great majority of rural poor are concentrated. An important impact of the strategies pursued in the 'Green Revolution' period has been intensification of regional disparities and imbalances in agricultural development and food availability and hence levels of food security. For the country as a whole while per capita availability of cereals has increased substantially, that of pulses has decreased significantly.

Table. Rice and Wheat Production

Production (million tonnes)

Year	Rice	Wheat	Foodgrains	R+W as % of Foodgrains
	million tonnes			
1970-71	40.80	20.86	108.4	57.4
1980-81	49.91	34.55	129.6	65.4
1990-91	72.78	53.03	176.4	75.0
1996-97	81.74	69.35	199.4	75.0
2010-11	82.00	72.00	205.4	76.0

In summary an annual increase in foodgrains production of 3.22 percent during fifties was mainly because of expansion in area. Sixties recorded a low annual growth rate of 1.72 percent necessitating large-scale imports of foodgrains. Annual growth of 2.08 percent was recorded during seventies. This decade was the turning point in India's foodgrains economy leading to self sufficiency through significant productivity increase first in wheat and later in rice in the eighties. An annual growth of 3.5 percent in foodgrains in eighties was the hallmark of green revolution that enabled India to become self-sufficient and even a marginal exporter. The pace of growth slowed in nineties barely making or even slower than the population growth rate. This is a matter of concern.

Table. Annual compound growth rate of foodgrains production

Crop	1950-51 to 1959-60	1960-61 to 1969-70	1970-71 to 1979-80	1980-81 to 1989-90	1990-91 to 1997-98	2010-11 to 2013-14
Rice	3.28	- 8.05	1.91	4.29	1.53	2.00
Wheat	4.51	5.90	4.69	4.24	3.67	3.90
Coarse cereals	2.75	1.48	0.74	0.74	- 0.49	-0.50
Total cereals	3.00	2.51	2.37	3.63	1.84	1.90
Pulses	2.72	1.35	- 0.54	2.78	0.76	0.90
Total foodgrains	3.22	1.72	2.08	3.54	1.66	1.68

Table. Per capita net availability of cereals and pulses per day (g)

Year	Cereals	Pulses	Total
1951	334	61	395
1961	400	69	469
1971	418	51	469
1981	417	38	455
1991	468	42	510
1998	451	33	484
2010	485	34	489

PRIMITIVE AGRICULTURE

Historical writers agree that in early days, the inhabitants of the earth gained a livelihood by hunting and fishing. When the population of a region became so dense as to deplete the supplies of food to be gained in this way, the more enterprising ones tamed and domesticated some of the wild animals for their own use. These animals were given protection and provided with food so that they multiplied rapidly. In this way a supply of meat was insured for those who cared for and protected them.

From hunting and fishing the people turned to such pursuits as grazing and herding, and gave much of their time to caring for animals. This was known as the pastoral stage of subsistence. It was more reliable than hunting and fishing for several reasons. By protecting the domesticated animals from the beasts of prey and driving away the less useful animals, larger numbers of useful animals were enabled to live in a given region. Food supplies of meat were therefore more plentiful and certain.

This custom of providing food supplies for animals led to the discovery that certain plants were better adapted for food than others. To raise these desirable plants in abundance it became necessary to destroy other plants that were not so useful, but which contended with the useful plants for possession of the soil.

This may be said to be the beginning of agriculture. At least it was the first indication of the organization of nature's forces to meet the needs of man. The demand for land for grazing and tilling purposes led to mutual agreements for territory which should be used for such purposes by certain persons or tribes, and is the first suggestion of land allotment or ownership.

BRITISH AGRICULTURE

The agriculture of England has passed through many different stages. Because of scarcity of agricultural laborers between 1350 and 1375, many of the owners of large estates were forced to change their system of farming. Previous to that time most of the tillable land had been devoted to grain raising, bread being the chief article of food. Then the land was changed into pasture, and sheep and cattle raising took a prominent place in British agriculture.

The greatest development came about the middle of the seventeenth century with the introduction o fclover and turnips as field crops. The chief advantage in these crops was the increased amount of feed, which enabled the farmer to keep more cattle on his land. Keeping of cattle gave manure that was applied to the land and resulted in better handling of the farms. The turnips took the place of bare fallowing which had previously been practiced, thus enabling the farmer to use all of his land each year.

Improved methods of tillage followed rapidly. About the first of the eighteenth century Jethro Tull started the practice of drilling the grain in rows, later advocating tillage between the rows, and inventing implements for the work. Lord Townshend, about the same time began the rotation of crops, greatly improving the producing power of his land thereby. Another notable man in British agriculture is Coke of Holkham, who like his predecessors grew turnips and clover, but who also studied economical management of his farm. He reduced the number of horses used in plow teams and increased greatly the efficiency of his live stock. Following the work of these three men, farming became a gentleman's profession in England and progress since has been rapid. The development and improvement of live stock has received so much attention at the hands of British breeders that the whole world practically has been drawn to their markets for foundation stock.

AMERICAN AGRICULTURE

The agriculture of America has been developed largely from the methods followed on the British Isles. The American colonists brought with them the customs of the British farmers and have since gone frequently to the mother country for improved seeds, for live stock, and even for men to do certain parts of the farm labour, as for instance, caring for the more valuable imported stock.

The development of agriculture during colonial days was slow and full of hardship. The British and European methods of agriculture required re-adapting to the new conditions. The people were forced to live a new kind of life, to clear land of trees and stones, and to learn by experience what crops were adapted and what methods of tillage would succeed. From the Indians they learned how to grow corn and tobacco, which have ever since been two of our important crops. From them also they learned many of the secrets of soil and climate. The agricultural development proceeded along two lines. In the South large plantations with negro slaves as laborers were established. Such establishments were practically self-sustaining because they supplied nearly all articles of food and clothing. In the North the family farm was developed largely and the labour of the farm was performed mainly by the farmer and his family. The conditions existing in the North encouraged the cooperative spirit and developed sociability among the people. Land in either section was abundant and the greatest problem in the management of farms waws that of economizing labour.

The colonial development was along the eastern sea coast. Following the Revolutionary war, settlement moved westward beyond the Alleghany Mountains and into the Great Central Basin. The invention of the cotton gin about 1793, and the development of the cotton industry, changed rapidly the systems of farming in the South. Slavery became greatly extended to provide cheap labour for the cotton fields, and probably had much to do with the rapid settlement of the southern states. Cotton and tobacco growing were the leading industries in these states and agricultural improvement was slow because labour was cheap as compared with that in the northern states and the use of improved machinery was not so imperative.

Migration westward was encouraged by the opening of the Northwest Territory in 1785. A liberal land policy on the part of the National Government and the discovery of gold in California served to stimulate the movement westward and aided the rapid settlement and development of the great prairie regions of the Mississippi Valley. The invention and perfection of farm machinery, begun about 1825 to 1830 and continued to the present time, has also been a great factor in the development of agriculture in America. The reaper, invented in 1831-1834, the thresher, perfected about 1850, and steel plows for turning the soil, manufactured since 1840, have had a great influence on the agriculture of the northern states which were especially interested in grain and corn raising.

Live stock raising was followed to some extent in the northern states also, but received great impetus about 1839 by the importation of the English thoroughbred stallion "Denmark" into Kentucky. Percheron draft horses from France were also introduced soon after. Corn raising on the western prairies stimulated greatly the growing of hogs. Cattle raising on the vast prairies also was greatly extended during this period. Dairying began to develop about 1850. The improvement of machinery and the increase in stock and grain raising went on steadily in the northern states, not being interrupted seriously even by the Civil War. In the South where cotton and tobacco grown on large plantations by slave labour were the main crops, agriculture was seriously demoralized by the war which set free the slaves, thus destroying the source of cheap labour. Several years were required to build up a new system of agricultural production and to adjust labour problems, but the agriculture of the South is now on a much better basis.

During the past twenty-five years agricultural development has been particularly rapid. The establishment of experiment stations in 1888 in every state in the Union, through the passage by Congress of the Hatch Act in 1887, and the development of Agricultural schools and colleges has led to a systematic application of science to agriculture. Crop growing has become much more certain and live stock growing has been largely developed. The occupation and settlement of the prairie lands, the perfection of machinery, and extended opportunities for agricultural education, have been large factors in improving conditions for farming. In the days of abundant land, large holdings were secured by many farmers. The tendency in later years is to reduce somwhat the size of farms and to take up more intensive forms of agriculture. Greater efficiency is demanded in the management of farms in order that greater profits may be realized by the tiller of the soil.

Farming Must Be Organized

During the period of cheap land and comparatively cheap labour, many crops were raised which do not under existing conditions, yield a profit. Methods of tilling the land and handling the crops that were followed during that period would not now give large enough yields to support the farmers and their families. Crops and methods of tillage must be readjusted to meet the changed economic conditions and satisfy the demands of the rapidly growing population for food stuffs. The practice of grain raising so prevalent in many portions of the United States must give way to live stock raising, fruit and vegetable growing and other more intensive systems of agriculture. A great variety of crops and a larger supply of all food stuffs and clothing material is required by the present population. The value of the agricultural products in the United States in 1910 was $8,694,000,000. The total value of all live stock on farms in the United States in 1910 was $4,925,000,000. The demand is constantly increasing. The future supply of these commodities is a matter demanding the most serious consideration.

It is important that the old soil robbing types of farming be superseded by more conservative methods; that the land be so tilled that it will bring the largest crops at the least expense of soil fertility; and that these crops be used or fed as near as possible to the land upon which they were grown, the manures and crop residues being returned to the soil. It is important also that the farmer receive a fair compensation for his labour. Crops must be selected that pay the best profits. These will differ in different years and localities. The farmer, therefore, must learn to select wisely and manage well so as to make a reasonable profit on his labour and investments.

The business of farming is constantly becoming more complex. The keen demand for land, the scarcity of labour, the desire for financial gain and the call of the city residents for supplies are acting as a spur for still greater effort. To meet these demands a new science is being developed which aims to correlate the various factors entering into the farmers' business so that it will be certain and remunerative. An effort is being made to organize the business of farming so that it may provide for the needs of the present generation without exhausting the source of supplies for generations yet to come.

AGRICULTURE IN THE CHANGING GLOBAL SCENARIO

Steady globalization of trade has profound implications for future agricultural development. The diversity of India's agro-ecological setting, high bio-diversity and relatively low cost of labour provide potential for agricultural competitiveness in a globalized economy. It is expected that with increasing globalization of markets over the years there will be demands for agricultural intensification. This will also be favoured because of greater backward and forward linkages between agriculture and food industry.

Therefore, increase in production and productivity are bound to be strategically important to economy. Intensification will not only favour alleviation of rural poverty but will also improve resource conservation particularly in the small farming sector where farmers can be encouraged to take up organized production of high value crops such as fruits, specialty vegetables, flowers medicinal and aromatic herbs etc. Stronger demands for crops of the small farmers' will not only improve incomes and welfare but will also make investments in technology and resource conservation more attractive. The General Agreement on Tariff and Trade (GATT) and liberalization of global trade is bound to have impact on future land use and production pattern. Understanding the local, national and international environment under which agricultural production is taking shape will be crucial in developing our own strategies.

EXTENSION STRATEGIES

Since early fifties a number of public by funded agricultural development programmes have been sponsored. These have included programmes like the National Extension Service (NES) Blocks in 1953, the Intensive Agricultural District Programme (IADP) in 1961-62, the Intensive Agricultural Area Programme (IAAP) 1964-65, the High Yielding Variety (HYV) programme 1966-67 and the Small and Marginal Farmers' Development Programmes (SMFDP) in 1969-70. Though these programmes had a perceptible impact the efforts did not get replicated over different areas and categories of farmers. In mid seventies based on pilot level project in Rajasthan Canal and Chambal command area a 'Training and Visit' (T&V) system of extension was promoted in different states. Extension efforts of the Indian Council of Agricultural Research through its research Institutes and the State Agricultural University were largely limited to demonstration of new technologies through such programmes as National Demonstration Project, Operational Research Project, the Lab to Land Programme and the Krishi Vigyan Kendras. However, there appears much to be desired in the way that extension programmes are conceived and implemented. At present extension programmes are implemented in largely a top-down fashion leaving little scope for localized planning and action. Farmers are almost passive receivers and their involvement in the process of technology generation and adoption is almost absent. Extension services, at present, are almost exclusively in the public sector domain and there is no effort or institutional support for other operators e.g. the NGOs, the corporate bodies etc. Extension programmes sponsored by the government operate largely in isolation and there appears a strong need to view the extension programmes as an integral part of the research and development process. The challenges facing agricultural development call for fundamental changes in our approach to technology transfer/extension programmes. Changes are necessary in the context of changing economic environment following policy adjustments in relation to privatization, deregulation and globalization calling for greater efficiency and effectiveness of the extension system.

More importantly there is need for

- Greater emphasis on providing producers with knowledge and understanding needed to overcome the problems or to exploit opportunities of their own specific production systems. Correspondingly there will be a need to de-emphasize 'package of practices' or the blanket recommendations, top down approach followed thus far.
- Shift in the focus of public extension systems from promoting inputs use to one on sustainable management of resources and improvements in the production system as a whole.
- Closer interaction between farmers, extension scientists and production system researchers in diagnosing problems and identifying location specific recommendations emphasizing participation and education rather than being prescriptive.
- Widening the range of extension delivering agencies. While the publicly operated extension systems will continue to be important, there will appear a greater role for NGOs, farmers' associations and corporate sectors in particular situations. Role of commercial suppliers of seeds, agrochemicals, machinery, vaccines and medicines in providing advisories, as is already being done in a limited way, will need to be encouraged and factored into public system's own priorities.
- Wider and more creative use of mass media in tune with current developments in information technology to get information across to the farming community whose ability to overcome constrains at farm level will increasingly depend on access to reliable and up-to-date information.

CHANGING LAND-USE AND FUTURE OF AGRICULTURE

One of the most important consequences of growing pressure on land is the declining trend in the average farm size and the pattern of holdings.

According to the latest Agricultural Census in 1970-71 there were 70 million holdings operating 162 million ha. By 1990-91 there were 105 million holdings operating 165 million ha. The average farm size decreased from 2.30 ha in 1970-71 to 1.57 ha in 1990-91. As of 1990-91 about 78 percent of holdings were small (1.0 to 2.0 ha) and marginal (<1.0 ha). A little more than 20 percent of the farmers were semi-medium (2.0 to 4.0 ha) and medium (4.0 to 10.0 ha). Large farmers (>10.0 ha) constituted only 1.6 percent of the total holdings. Over the twenty-year period since 1970 the proportion of marginal farmer has increased from 50 to 59 percent and that of large farmer has declined from about 4 to 1.6 percent.

Table. India – Land Use Categories (million ha)

Classification	1960-61	1970-71	1980-81	1990-91	1993-94
Reporting area	298.5	305.0	305.0	305.0	305.0
Area under non-agricultural uses	14.8	16.5	19.7	21.1	22.0
Barren and un-cultivable	35.9	28.2	20.0	19.4	19.0
Cultivable waste	19.2	17.5	16.7	15.0	14.5
Old fallow	11.2	8.8	9.9	9.7	9.7
Forests	54.1	63.9	67.5	67.8	68.4
Permanent pastures	14.0	13.3	12.0	11.8	11.2
Land under misc. trees etc.	4.5	4.3	3.6	3.8	3.7
Net sown area	133.2	140.3	140.0	143.0	142.1
Current fallow	11.6	11.1	14.8	13.7	14.3

The proportion of total area operated by marginal farmers increased from nine percent in 1970-71 to nearly 15 percent in 1990-91 while the proportion of large farmers declined from about 31 percent to 17 percent in the same period. The size of average holding is very unevenly distributed among the states. States with relatively large average size of operational holding are a mixed lot – they include states with large tracts of barren lands e.g. Rajasthan, Maharashtra and Madhya Pradesh on the one hand and agriculturally advanced state like Punjab on the other. These trends in farm size changes will have a profound effect on the future agricultural development strategies.

Table. Livestock and poultry population (million)

Category	1982	1987	1992
Cattle	192.45	199.69	204.53
Buffaloes	69.78	75.96	83.49
Sheep	48.76	45.70	50.79
Goats	95.25	110.20	115.28
Total (Livestock)	419.59	445.28	470.15
Total Poultry	207.74	275.32	307.10

Although India's population growth rate has slowed from 2.1 percent in 1980s to 1.8 percent in the 1990s and is expected to slow further in the coming decades, yet the population is projected to reach 1.33 billion by 2020 from the current one billion. The urban share of total population is projected to increase from 26 percent to 35 percent of the total population. Although incidence of poverty is falling, it is estimated that in 93-94 (upto which data is available) 320 million people constituting 36 percent of the population were below the officially defined poverty line.

The nature of the poverty line has been shifting. About 30 years ago 48.4 percent of those living in rural areas were poor and 20 percent of those living in the urban areas were classed as poor. Recent studies show that the number of poor in urban areas have been increasing at relatively higher rate compared to the rural areas. At present those below the poverty line in rural sector constitute 37 percent of the population while in the urban sector the percentage is 32 percent. In the context of poverty alleviation, therefore, emphasis will be required to be placed both on production of food by the poor as well as on the availability of food for the urban poor.

It needs to be recognized that a large proportion of the rural poor are located in regions of low potential for food production e.g. arid and semi-arid areas, hilly regions, degraded land and forest areas. Widespread hunger and malnutrition are the direct manifestation of poverty and will call for increasing efforts to produce more food at affordable price. Increasing population and economic growth are changing patterns of land use making potentially unsustainable demands on the country's natural resources.

Since early fifties the net area sown was expanded rapidly at first but at a diminishing rate since 1970 to reach approximately 142 million ha at present.

During 1950s and 1960s areas under agriculture expanded substantially as the fallows were reduced and cultivable wastes were put under the plough. The net area sown increased from 119 million ha in 1950-51 to 133 million ha by 1960-61 and further to 140 million ha by 1970-71. Fallow lands declined from 28 million ha in 1950-51 to 20 million ha by 1970-71. Cultivable wastelands declined from 23 to 17.5 million ha.

Land use intensity i.e. fraction of net sown area to total geographical area increase from 36 percent in 1950-51 to 40.5 percent in 1960-61 and 43 percent by 1970-71 where it has since stabilized.

- Cropping intensity i.e. gross sown area as percent of net sown area increased from 111 percent in 1950-51 to 115 percent in 1960-61, 118 percent in 1970—71 and 130 percent by mid 1990s.

- While the contribution of increased area in the growth of agriculture has declined over time, that of productivity has increased. The yield of all crops grew at 1.5 percent per annum between early 1950s and mid 1960s. the pace accelerated to 1.7 percent in the 1970s and then to 3 percent per annum between early 1980s and mid 90s. Unlike the gains in area, which benefited non-foodgrains, the gains in productivity accrued mostly to foodgrains.
- India's forest resources have been dwindling. According to the 'State of Forest Report' (1997) the total forest cover of the country is estimated at 63.34 million ha i.e. 19.27 percent of the geographic area of the country. Of these the dense forest (crown density more than forty percent) and open forest (crown density 10 to 40 percent) occupying about 11 and 8 percent of the geographic area respectively and mangroves occupy 0.15 percent of the geographic area. The country has lost about 5482 sq. km. of forest cover since the 1995 assessment. By any estimate the area under forest is far below the national policy goals and many areas nominally under forest are being used for non-forest purposes. Similarly 'uncultivated lands' such as permanent pastures, miscellaneous tree crops, cultivable wastes and fallow is subject to increasing competition from uses other than feeding livestock.
- The growth of livestock population is an important source of competition for land. The increase in number of major classes of livestock.
- The area sown to fodder crops is not recorded. Information available from other sources provide an estimate ranging from 4 to 5.5 percent of the net sown area and suggest that the area under fodder crops will have to increase to 10 percent or more to support increasing livestock based activity.

The pressure on India's land and water resources is seriously threatening native plant and animal diversity. India has uniquely rich and diverse genetic base. With increasing agriculture and economic development the genetic pool is declining. This decline, if unchecked and poorly managed can have unforeseen and adverse consequences for the sustainability of agriculture of the region.

TECHNOLOGICAL NEEDS AND FUTURE AGRICULTURE

It is apparent that the tasks of meeting the consumption needs of the projected population are going to be more difficult given the higher productivity base than in 1960s. There is also a growing realization that previous strategies of generating and promoting technologies have contributed to serious and widespread problems of environmental and natural resource degradation. This implies that in future the technologies that are developed and promoted must result not only in increased productivity level but also ensure that the quality of natural resource base is preserved and enhanced. In short, they lead to sustainable improvements in agricultural production.

Productivity gains during the 'Green Revolution' era were largely confined to relatively well endowed areas. Given the wide range of agroecological setting and producers, Indian agriculture is faced with a great diversity of needs, opportunities and prospects. Future growth needs to be more rapid, more widely distributed and better targeted. Responding to these challenges will call for more efficient and sustainable use of increasingly scarce land water and germplasm resources.

Technical solutions required to solve problems will be increasingly location-specific and matched to the huge agroecological/climatic diversity. Detailed indigenous knowledge and greater skills in blending modern and traditional technologies to enhance productive efficiency will be more than ever before, key to the farming success and sectoral growth. Most technological solutions will have to be generated and adapted locally to make them compatible with socio-economic conditions of farming community.

New technologies are needed to push the yield frontiers further, utilize inputs more efficiently and diversify to more sustainable and higher value cropping patterns. These are all knowledge intensive technologies that require both a strong research and extension system and skilled farmers but also a reinvigorated interface where the emphasis is on mutual exchange of information bringing advantages to all. At the same time potential of less favoured areas must be better exploited to meet the targets of growth and poverty alleviation.

These challenges have profound implications for products of agricultural research. The way they are transferred to the farmers and indeed the way research is organized and conducted. One thing is, however, clear – the new generation of technologies will have to be much more site specific, based on high quality science and a heightened opportunity for end user participation in the identification of targets. These must be not only aimed at increasing farmers' technical knowledge and understanding of science based agriculture but also taking advantage of opportunities for full integration with indigenous knowledge. It will also need to take on the challenges of incorporating the socio-economic context and role of markets.

With the passage of time and accelerated by macro-economic reforms undertaken in recent years, the Institutional arrangements as well as the mode of functions of bodies responsible for providing technical underpinning to agricultural growth are proving increasingly inadequate. Changes are needed urgently to respond to new demands for agricultural technologies from several directions. Increasing pressure to maintain and enhance the integrity of degrading natural resources, changes in demands and opportunities arising from economic liberalization, unprecedented opportunities arising from advances in biotechnology, information revolution and most importantly the need and urgency to reach the poor and disadvantaged who have been by passed by the green revolution technologies.

Another important implication of increasing globalization relates to the need for greater attention to the quality of produce and products both for the domestic and the foreign markets. This would imply that production must be tuned to actual rapidly changing product demand. Such adaptation to global markets would require state of the art research, which can be achieved only by setting global standards of research, focus on well defined priorities and mechanisms which permit close interaction of farmers with researchers, the private sector and markets.

2

Agricultural Land and Farm Management

The land surface of the United States covers 2.3 billion acres. Private owners held 61 percent in 2002, the Federal Government 28 percent, State and local governments 9 percent, and Indian reservations 3 percent. Virtually all cropland is privately owned, as is three-fifths of grassland pasture and range and over half of forestland. Federal, State, and local government holdings consist primarily of forestland, rangeland, and other land. Most land in Federal ownership—largely in the West—is managed by the Department of the Interior (68 percent) and the Department of Agriculture (28 percent) (U.S. GSA, 2005).

Farm operators do not own all the land used in agriculture. According to the 1999 Agricultural Economics and Land Ownership Survey (AELOS), farmers held 58 percent of the land in farms in 1999 (USDA, 2001). These landowning farmers also made up 58 percent of the 3.4 million farmland owners.

Nonoperator landlords accounted for the remaining 42 percent of land in farms. Ninety-five percent of nonfarm landlords were individuals/families or partnerships. Of these unincorporated landlords, 55 percent were at least 65 years old. Many nonfarm landlords have a historic connection to farming. Among the people who have exited farming or inherited farmland since the number of farms peaked during the Great Depression, a number have retained ownership of some or all their land.

LAND POLICY

Land policy in India has been a major topic of government policy discussions since the time prior to Independence from British rule. The peasants of the country strongly backed the independence movement and the "Land to the Tiller" policy of the Congress Party because of the prevailing agrarian conditions. The agrarian structure during British administration emerged with a strong historical background (Baden Powel, 1974; Dutt, 1976; Appu, 1996). The land-revenue system implemented by Todar Mal during Akbar's regime can be traced as the possible beginning of systematic efforts to manage the land. This method incorporated measurement, classification and fixation of rent as its main components. Under the various pre- British regimes, land revenues collected by the state confirmed its right to land produce, and that it was the sole owner of the land. British rulers took a cue from this system and allowed the existence of noncultivating intermediaries. The existence of these parasitic intermediaries served as an economic instrument to extract high revenues (Dutt, 1947) as well as sustaining the political hold on the country. Thus at the time of Independence the agrarian structure was characterized by parasitic, rent-seeking intermediaries, different land revenue and ownership systems across regions, small numbers of land holders holding a large share of the land, a high density of tenant cultivators, many of whom had insecure tenancy, and exploitative production relations (Appu, 1996).

Immediately after Independence a Committee, under the Chairmanship of the late Shri J. C. Kumarappa (a senior Congress leader), was appointed to look into the problem of land. The Kumarappa Committee's report recommended comprehensive agrarian reform measures. India's land policy in the decades immediately following its independence was dominated by legislative efforts to address the problems identified by the Kumarappa Committee . A substantial volume of legislation was adopted, much of it flawed and little of it seriously implemented.

Several important issues confronted the policy-makers.

1. Land was concentrated in the hands of a few and there was a proliferation of intermediaries who had no vested interest in self-cultivation. Leasing out land was a common practice.
2. The tenancy contracts were expropriative in nature and tenant exploitation was ubiquitous.
3. Land records were in extremely bad shape giving rise to a mass of litigation. It is ironic that the Supreme Court of India in 1989 commented that the revenue records are not legal documents of title (Wadhwa, 1989). This is a sad commentary on the land records of the country.

It is against this background that land policy has been shaped in India. While land-reform legislation remained active, land policies in more recent decades have focused less on land reform and more on land development and administration.

Land policy in India has undergone broadly four phases since Independence.

1. The first and longest phase (1950 - 72) consisted of land reforms that included three major efforts: abolition of the intermediaries, tenancy reform, and the redistribution of land using land ceilings. The abolition of intermediaries was relatively successful, but tenancy reform and land ceilings met with less success.
2. The second phase (1972 - 85) shifted attention to bringing uncultivated land under cultivation.
3. The third phase (1985 - 95) increased attention towards water and soil conservation through the Watershed Development, Drought-Prone Area Development (DPAP) and Desert-Area Development Programmes (DADP). A central government Waste land Development Agency was established to focus on wasteland and degraded land. Some of the land policy from this phase continued beyond its final year.
4. The fourth and current phase of policy (1995 onwards) centres on debates about the necessity to continue with land legislation and efforts to improve land revenue administration and, in particular, clarity in land records.

This paper is an attempt to discuss the critical issues pertaining to land policy in India beginning with the emergence of a post-Independence policy in a historical context, and from a viewpoint of differential provincial policies. Naturally, land reforms predominate the discussion here. Land reforms have been one of the important land policy initiatives in India that have brought a fundamental change in the entire approach towards development. The paper discusses the impact of land reforms and the changing phases of land administration. The focus is on the role and development process of land policy in India in the context of overall changes in India's development policies. It also addresses the political and economic aspects of the policy initiatives, beginning with the various land-reform efforts and finally analysing the recent land development and administration policies. The paper incorporates a discussion on the closely related goals of land policy, i.e. poverty elimination, conflict management, sustainable economic growth and good environmental management. In the final analysis, the paper highlights current issues pertaining to the relationships among land policy, poverty and the development initiatives.

EMERGENCE OF LAND POLICY IN INDIA

Given India's vastness, diversity and various political, economic and social influences from a history of various rulers and foreign conquerors, it is not surprising that land tenurial and administration practices varied significantly throughout the subcontinent at the time of Independence. One common factor was that land policies had been driven by the rulers' efforts to extract land revenue or tax from those working on the land. Throughout much of the country, the rulers appointed Zamindars, or tax collectors, who were contracted to collect land revenue for a given large territory and pay fixed sums to the government (but often extracted as much as they could from the landholders and pocketed the difference). Though tenurial conditions varied significantly from region to region, the numerous tenures could be classified under two broad categories - the Zamindari and Rayatwari systems. The Zamindari system was characterized by one or more layers of proprietary rights between the state and the actual landholder. In the Rayatwari (or peasant proprietorship) system, no intermediaries existed in design but emerged in the process.

The British rulers continued with existing land-revenue policies and procedures with a few but significant modifications. Perhaps most importantly, the British made the tax-collecting Zamindars into proprietors of the estates over which they had tax collection duties. This change was aimed at accomplishing two objectives: simplifying the land-revenue collection process and creating a rural elite with a vested interest in British rule. Unfortunately, it converted the erstwhile landowners into insecure tenants. Over time, many Zamindars assigned their land-revenue collection duties to one or more layers of intermediaries who were also given interests in the land. The historical emergence and perpetuation of intermediaries served the purpose of land revenue administration and political control of the successive rulers, but their numbers swelled. The large patches of land held by them were let to tenants at exorbitantly high rents. That created a disincentive among the tenant cultivators to develop the land, and consequently impacted upon production. Thus, the Colonial Government, out of its interest to administer the country effectively, did not make any substantial changes in the land-revenue system but promoted the class of non-cultivating intermediaries.

At the time of Independence, India faced a major challenge of setting right the agrarian structure as promised during the independence struggle. Thorner and Thorner (1961), in an analysis of the agrarian structure of India, vividly describe the pre-Independence structure as a complex of legal, economic and social relations - a multilayered structure that pulled down the production efficiency in the agricultural sector. A brief review of the literature also reveals a myriad of agrarian relations in India, varying from peasant proprietorship to a pure landlord - serf relationship. The first task placed before the first Indian parliament was to address land policy. Because India has a densely populated agrarian economy, almost all other developmental initiatives also involved land as a central and a complex issue, as it clearly represented social status and not just the means of production.

While recognizing the need to bring about land reforms in the country, the Constitution of India provided under Article 39 that: (1) the ownership and control of the material resources of the country should be so distributed as best to serve the common good; and (2) the operation of the economic system should not result in a concentration of wealth or a means to production to the common detriment.

The Constitution of India also made land a state (provincial) subject. So, only state (provincial) legislatures have the power to enact and implement land-reform laws. However, the central government played a significant advisory and financial role in land policy based on its constitutional role in social and economic planning (a role held concurrently with the states). The Government of India established a National Planning Commission immediately after Independence to fulfil this role of social and economic planning.

The Planning Commission has prepared a series of Five-Year Plans since 1951. Land policy has been one of the important components incorporated in all the plans. The policy statements are sometimes quite explicit in the plan documents, but are more often implicitly stated. An overview of changes in the land policy as reflected through the various plan documents. Land reform policy was spelt out in the First Five-Year Plan. The plan aimed to reduce disparities in income and wealth, to eliminate exploitation and to provide security to tenants, as well as to achieve social transformation through equality of status and an opportunity for different sections of the population to participate in development initiatives.

LAND REFORMS

Despite attempts at land reforms over successive Plan periods, the basic character of the agrarian economy has not undergone any structural change. The pattern of land distribution is highly skewed, with a high concentration of land in the hands of a few land owners on the one hand and the growing number of marginal and sub-marginal farmers on the other. Fragmentation of land holdings continues on a large scale and only a few States like Punjab, Haryana, Uttar Pradesh and parts of Maharashtra have been able to successfully undertake a programme of consolidation of holdings.

Agricultural tenancy, which was abolished in most of the States by various enactments in the post-Independence era, continues unabated though it is largely concealed. In the wake of liberalisation, several State Governments have modified their land ceiling laws so as to exempt orchards, fish ponds etc., from the purview of land ceilings. There is also a move to make suitable changes in tenancy regulations to attract private corporate investment in agriculture. Hence, it is necessary to reconsider the issue of land reforms, particularly from the point of view of the poor, as access to land is still a major source of livelihood in rural India. In fact, it has been argued that the need for poverty alleviation programmes has arisen because the land reforms have not been implemented in a systematic way. The experience of several countries in East Asia shows that land reforms, leading to structural equity in the distribution of land, are an essential prerequisite for economic development through agricultural transformation. In addition, the efficiency of land use and land management, and protection of the land rights of the tribals and women have assumed great significance in the context of the changes that are taking place in rural India.

The continued importance of land reforms was recognised in the Eighth Plan, with the abolition of intermediaries, redistribution of ceiling surplus land, tenancy reforms providing security of tenure to tenants and share croppers, consolidation of holdings and updating of land records as the main objectives of the land reform policy. However, only limited success was achieved with respect to these objectives in the Eighth Plan. Given that land reforms is a State subject, the Central Government can only draw the attention of the State Governments to the pressing needs for land reforms, which are central to any strategy of poverty alleviation.

At the end of the Seventh Plan, out of the 72.2 lakh acres of land declared surplus, only 46.5 lakh acres had been distributed. At the end of the Eighth Plan, out of the total 74.94 lakh acres declared surplus, 52.13 lakh acres had been distributed. In other words, during the Eighth Plan only 6-7 lakh acres were redistributed. Further, 12.4 lakh acres were under disputes pending in courts and 19.59 lakh acres were not available for distribution because they were unfit for cultivation or reserved for public purposes or for other miscellaneous reasons. In fact, only 59,000 acres were available for redistribution. Of the Bhoodan land donated, 53 per cent was distributed, accounting for 24.52 lakh acres. In addition, 142.87 lakh acres of wastelands were distributed among 88.5 lakh beneficiaries. But, there is still considerable scope for redistributing Government wastelands, common lands, ceiling surplus land and Bhoodan land.

Similarly, in the area of tenancy reforms very little progress has been made, after the initial abolition of 'zamindari' and the transfer of title to owner-cultivators in the immediate post-Independence period. The successful implementation of tenancy laws has been confined to West Bengal, Karnataka and Kerala. In fact, in the Eighth Plan there was no progress in respect of conferment of rights on tenants and therefore the issue of tenancy reforms is still illusory, but requires tackling.

Consolidation of holdings has taken place in very few States. While 15 States had enacted appropriate legislation, Andhra Pradesh, Tamil Nadu, Kerala, Pondicherry and the North-Eastern States do not have any laws for consolidations of holdings. Several States like Bihar, Maharashtra and Rajasthan have suspended the programme. In fact, only in Uttar Pradesh, 900-1000 villages are being covered annually.

There is evidence of considerable alienation of tribals from their land. As per the latest available estimates, 4.6 lakh cases of tribal land alienation covering 9.2 lakh acres have been registered. Of these 2.7 lakh cases covering 6.3 lakh acres have been disposed of in favour of tribals but physically an estimated 4.7 lakh acres had been restored to them. In other cases, reconciliations are being effected.

It cannot be gainsaid that the essential prerequisite of any land reform measure is the recording of land rights and their updating. In recognition of this, a Centrally Sponsored Scheme for Strengthening of Revenue Administration and Updating of Land Records was introduced during the later half of the Seventh Plan and against an outlay of Rs.20.8 crore, Rs.14 crore were spent. In the Eighth Plan a provision of Rs.175 crore was made against which Rs.98 crore were released and Rs.66.7 crore utilised.

But, given that the funds are shared in the ratio of 50:50 between Centre and States, several States have not been able to provide their share and hence, the utilisation has been low. Several other States have not availed of this scheme at all. In 1995-96, funds were released only to Bihar, Kerala, Maharashtra, Madhya Pradesh, Rajasthan, Tamil Nadu and West Bengal aggregating to only Rs.18.8 crore.

The States have been requested repeatedly to expedite expenditure and to adopt new technologies for survey and settlement operations, preparation of maps etc. Funds for infrastructure development were also sanctioned to meet the training needs of the revenue functionaries. Some States have developed new training institutions, while a few have upgraded the existing ones. In 1988-89, 8 pilot projects were taken up for computerisation of land records with 100 per cent Central assistance. Since inception, 323 projects have been sanctioned and funds to the tune of Rs.64.43 crore have been released. However, the utilisation was only to the extent of 20 per cent. This shows the tardy progress made in the implementation of this scheme on the ground.

Land Reforms in Ninth Plan

In the Ninth Plan the issue of agrarian restructuring will continue to receive the top most priority in the expectation that the States would be able to facilitate changes that would make for more efficient agriculture, leading to increases in both output and employment. This process will, in turn, contribute to the achievement of a higher rate of economic growth with social justice.

The main components of the land reform policy are the detection of ceiling surplus land and the distribution of the existing surplus land, besides tenancy reform, consolidation of holdings, providing access to the poor on common lands and wastelands, preventing the alienation of tribal lands and providing land rights to women. However, for the successful implementation of land reforms, updating of the land records, both by traditional methods and through computerisation, is an essential prerequisite. Let us elaborate.

Ceiling on Land Holdings: With the introduction of the Land Ceilings Act in 1972, the ceiling on land holdings was introduced in almost all the States with the exception of some North-Eastern States, though the ceiling limit varied depending on the quality of the land. The ceiling surplus land was to be distributed among the landless poor. In this way, land ceiling was considered an important instrument for reducing disparities in the ownership of land and as a way of increasing productivity through greater utilisation of labour. However, in practice, the extent of land declared surplus was very limited. The total area declared surplus is estimated at less than 2 per cent of the total cultivated area. Further, distribution of surplus land has been limited in several States because of institutional and legal rigidities. There are other reasons too which have led to the poor performance with regard to land ceiling. Most importantly, the ceiling surplus land continues to exist in a concealed way particularly in areas which have been covered by irrigation in the post-1972 era. Further evidence suggests that there are areas where land owners have land in excess of ceiling limits which can be mopped up if the programme for unearthing it is pursued vigorously. Also, in order to circumvent the ceiling laws, benami and furzi transfers are effected which need to be identified. There is clearly a need to detect the land falling within the ceiling limits and redistribute it. But detection and distribution of ceiling surplus land require tremendous political and administrative will.

In the wake of the economic liberalisation in certain States like Karnataka, the industry and the large farmers are being given exemption from ceiling laws without seeking the permission of the Government of India. This would certainly go against the interest of the poor as it would increase landlessness and depress agricultural wages. Hence, this issue requires close examination, before such exemptions are given.

Small farms are efficient in terms of yields and returns per unit of area. However, they are often not able to generate adequate income to sustain further investment. Therefore, there is a need for both horizontal and vertical diversification of small farms. An analysis of Indian Council of Agricultural Research (ICAR's) data of national demonstrations shows that there are large technological/yield gaps, particularly in the backward regions, which, if fully exploited, would enable the marginal and small farmers to generate incomes well above the sustenance level, subject to crop rotation, cropping pattern, yield responses of technological adoption, etc. In fact, in irrigated areas, the small and marginal farmers are viable because of their capacity to produce two or three crops a year. In the case of rainfed agriculture, a breakthrough in dryland farming is required with respect to the traditional cereal crop that can grow in these areas.

Alternatively, there is a possibility of shifting to other high-value crops like fruits, vegetables, mulberry, etc. Other support systems would have to be appropriately developed to provide facilities for credit, marketing, storage and transportation of the perishable high-value crops. In fact, a change in the cropping pattern away from cereals to some high-value crops, would certainly make the small and marginal farmers viable. Also, given the increasing market demand for certain products, sericulture, horticulture and aquaculture have a great potential. In Maharashtra and West Bengal a small piece of land yields adequate returns from horticulture and acquaculture respectively. This provides further justification for ensuring that ceilings are strictly enforced as small farms can be perfectly viable in the given context.

As observed in the UNDP Human Development Report 1996, as land is redistributed from big to small farms, not only the family labour per hectare can increase sharply, so, can hired labour also. For both the reasons, the employment situation improves even for those, who remain landless after the land reform.

The main conclusion from this is that an agricultural strategy centered on small farms, rather than large, simultaneously increases the social efficiency of resource use in agriculture and improves social equity through employment creation and more equal income distribution, that small farms generate.

While there appears to be a rationale for reducing the existing ceiling limits further with a view to alleviating poverty of the growing number of rural landless poor, this is unlikely to have the support of the State Governments.

Therefore, a pragmatic approach would be to strictly enforce the existing ceiling limit without permitting any attempt to circumvent it. Given the employment elasticity of agriculture in areas, which are agriculturally backward, land reforms would ensure both agricultural growth and greater employment for the rural poor. It would also provide a social status to the large number of poor. A greater transparency in the method of distribution of surplus land is possible with a greater involvement of panchayats, local communities and NGOs.

Tenancy Reforms: The policy with regard to tenancy clearly provided for conferment of ownership rights on tenants or for acquisition of ownership rights by them on payment of a reasonable compensation to the landlords. The tenancy laws of most States abolished tenancies so as to vest ownership of land with the actual tiller. Despite the legal provisions to abolish tenancy, it has continued to flourish with the existence of a large number of tenants and sub-tenants without any protection against eviction leading to insecurity among them. The continued existence of oral and concealed tenancies has led to low investments and low productivity in agriculture in several States where the implementation of tenancy reforms has been tardy. In contrast, in West Bengal, Karnataka, and Kerala, much success has been achieved in this respect. The success of 'Operation Barga' in West Bengal is well documented. This shows that conferring occupancy rights on the tenants has led to better investments in land and consequently, a higher rate of return. Hence, it is desirable that States which have a high incidence of concealed tenancy recognise this fact and record the rights of tenants. In addition, absentee landlordism has to be restrained. To the extent that large land owners leave their land fallow, either the State should take it over and lease it out on a long-term basis or the land owners should be required to lease it out on a long-term basis. Measures would have to be taken for the protection of tenants against displacement, eviction and other forms of exploitation.

A ban on tenancy was imposed in almost all the States to encourage owner cultivation and to give security of tenure to the sharecroppers and the tenants. In areas characterised by semi-feudal modes of production, where agricultural markets are not well developed, this ban is desirable with a view to protecting the tenants. It has been demonstrated that in 1972-73 in Purnea district of Bihar 40 per cent of the land was under sharecropping. This percentage has gone down and is probably around 25 per cent. The share of landholders is limited to 25 per cent of the gross produce but in practice this is perhaps illusory. Since the tenancies are oral and the sharecroppers are weak their hold on land is tenuous and they have to give to the landlord more than half of the produce.

In this case, the tenants have no incentive to make long-term investments in land. The landlords also do not look upon it as a productive asset but a store of value and a reflection of their social status. Both the landlord and the sharecropper would gain if the sharecropper had the rights of cultivation in the land as he would make greater investments which would lead to higher returns. This shows that in areas where tenancy flourishes, reforms are required together with a proper implementation of laws.

On the other hand, in the "Green Revolution" areas where there is greater awareness and the markets work, freeing the lease market for land would contribute both to equity and efficiency. In these areas where the traditional arrangements exist, the tenants have been reduced to the status of an agricultural labourer with the landlord exercising considerable influence. In this case, the tenants as well as small and marginal farmers would be able to augment their operational holdings by leasing in land and, with a greater intensity of cultivation, it would lead to greater output. Clearly, agricultural tenancy should be opened up and leasing in of land permitted subject to the ceiling limits. This would activise the land market which would enhance the poor people's access to land.

Protection of Tribal Land: Despite the commitment that the tribal lands must remain with the tribals, alienation of the tribals from their land continues on a large scale due to various legal loopholes and administrative lapses. Hence, in the Ninth Plan legal provisions must be made for the prevention of alienation of tribal lands and for their restoration, not only in the notified scheduled tribe areas but also in the tribal lands in other areas. Also, the regulation of resale of the tribal lands should be made as stringent as possible. Given that alienation is basically a consequence of economic deprivation and social discrimination it is felt, that for the tribal communities, various development programmes must be dovetailed in order to improve the income level of these people and to provide them with a basis for sustained livelihood. Even in the case of acquisition of lands of the tribals by the Government for public purposes, it should be restricted to a minimum. Encroachment of land belonging to tribals should not be permitted. Even civil courts should have no jurisdiction in the proceedings involving transfer of land of persons belonging to scheduled tribes.

Consolidation of Holdings: As already stated, consolidation of holdings has been successful in a very few States though several States have enacted legislation in this regard. In some cases, there are genuine problems in the process of consolidation including proper valuation of land, fear of eviction of small and marginal farmers who are tenants, inadequate availability of staff and lack of updated land records. Despite these constraints, consolidation of holdings makes for efficient land use and water management, leading to higher productivity and, therefore, must be enforced wherever practicable. However, in so far as there is insecurity among the tenants and small and marginal farmers, the State must ensure that their interests are protected. Consolidation operations, whenever undertaken, should be integrated with survey and settlement operations in order to avoid duplication in work and harassment to the affected person. The involvement of the Panchayati Raj Institutions should facilitate this process through a greater participation of the village people.

Government Land and Common Property Resources: In rural areas every village has common lands as well as other common property resources. These are a source of sustenance to the landless. Evidence suggests that considerable area of the government land has been taken over both by the rural poor as well as by the rural elite for agricultural and housing purposes. Unluckily, due to negligence, unsustainable overuse and excessive pressure of population on land, the productive capacity of common lands has been diminishing. However, joint forest management and watershed development programmes are schemes which can be successful in the regeneration of these common lands.

Clearly, 'pattas' should be given to the rural poor in order to provide them access to a means of livelihood. There are several success stories, both in the area of joint forest management as well as watershed development by groups of people, especially women on common land, which could be replicated. These lands can also be used for providing grazing land, fodder and fuel to the poor. If the common property resources vest in the Gram Panchayat, access to it should be limited to the rural poor and the rural elite should not be allowed to encroach upon it. Similarly, the wastelands which are lying unutilised should be reclaimed and distributed among the marginal farmers and landless labourers of the area. This would provide them a source of income as also generate greater output from a variety of activities in the area of agriculture, horticulture, fodder, fuelwood and other agro-forestry. Given the paucity of 'cultivable' land, it is necessary to redistribute the wastelands and fallow land in order to provide productive employment to the poor.

In the case of land vested and in possession of the Government and/ or where a final decision is pending in a judicial court, it is suggested that such land should not be kept vacant but allotted to eligible rural poor on a short-term basis with a clear understanding that they would have to handover the land to the original landlord if the court so directs. However, legal opinion is being sought on the possibility of taking up this measure. Again in the case of Government wasteland and degraded forests, there is a growing opinion that such land should be allotted/leased to the corporate sector and for industrial and commercial purposes. As argued above, this too would adversely affect the rural poor. Access to such land should be restricted to the rural poor in order to provide them with employment.

Gender and Land Rights: So far, the strategy of land reforms has not given any cognizance to the existing gender inequalities in land inheritance laws and ceiling laws. In most regions of the country, women constitute a disproportionate number of poor. They are also more dependent on agriculture for a livelihood than men, as men shift to non-farm employment. Also, it is estimated that 20 per cent of the rural households are *de facto* female-headed. Yet, very few women have titles to land and even fewer control it. Hence, ensuring women's effective command over land will be one of the new priorities of the Ninth Plan.

Traditionally, it was accepted that agricultural land would be inherited by sons, even though in some States the inheritance law did not stipulate such a provision. Hence, it is necessary to implement the laws and also to record the rights of women, where it is legally claimed by them. Further, in so far as inheritance rights in tenancy laws are subject to specific land acts, the issues cannot be easily resolved. In some States the tenancy laws provide for devolution of tenancy to male descendants and only in their absence, women can inherit, but then also to a limited extent. Hence, it is recommended that States should amend their land and inheritance laws so as to bring agricultural land at par with other forms of property. There has been some progress in this direction in Southern and Central India.

Many of our States have improved women's access to land and landed property. States like Karnataka, Tamil Nadu and Andhra Pradesh have amended the Hindu Succession Act, 1956 to formalise issues related to women's right to property including land. Some, like Rajasthan and Madhya Pradesh have provided that issues related to property, including landed property, would be dealt with in accordance with the appropriate personal laws. However, serious anomalies continue to persist. A number of States like U.P. Haryana, J&K, Delhi and Punjab are apparently yet to take adequate steps to provide the Constitutional/legal safeguards to women with respect to their access to land. Additionally, it is necessary that when ceilings are fixed in relation to a "family unit", the definition of a family should be uniform across States and sons and daughters should be given equal consideration, both while assessing ceiling surplus land and in land distribution under various resettlement programmes. The pattas should be in the name of women to a larger extent. While joint pattas are better than no pattas, which do not provide the women control over it. In fact, groups of poor rural women should be given group pattas with usufructuary rights, but no right to sell individually. This group approach would enable the women to retain control over land. They could invest in the land collectively and cooperate in sharing both labour time and the returns. There are several instances of such joint management by groups of women which need to be replicated. However, to enable women to reap the benefits of land acquisition, greater access to information, credit, inputs, marketing and technologies must be provided to them.

In order to have a better data base, both the Agricultural Census and the National Sample Surveys should provide information on land ownership, land operated and land under tenancy by gender. So far, land-based statistics are recorded on a household basis. A pilot survey should be undertaken proceeding the next nationwide NSS data collection exercise.

Updating of Access to Land Records: Maintenance of up-to-date land records is crucial for effective implementation of land reforms. In recognition of this, during the Seventh Plan period, a Centrally Sponsored Scheme for Strengthening of Revenue Administration and Updating of Land Records was introduced with a view to strengthening survey and settlement organisations at the grassroots level through training, equipment, staff etc. Hence, it is imperative that States should give special emphasis to resurveys and adopt appropriate modern technologies for this, including aerial survey, photogrametic systems, global positioning system, use of scanners, digital computerised maps etc.

Further, a special drive for recording land rights of tenants and sharecroppers, as in West Bengal, must be undertaken by other States. As recommended by the Conference of Revenue Ministers held in 1997 all the existing tenancy reform legislations should be examined and a model legislation prepared. Another Centrally Sponsored Scheme for Computerisation of Land Records was started during 1988-89 as a pilot project. But the scheme has made little progress. In fact, several factors including power shortage, lack of trained staff and bureaucratic procedures involved at various stages, have constrained the progress of computerisation of land records. Despite these problems, computerisation must be given a high priority and solutions found for the operational problems.

The updating of land records can be expedited even without computerisation through the involvement of panchayat institutions and the local revenue functionaries. The village-level revenue functionaries should be placed under the control of the Gram Panchayats, though the appellate jurisdiction should continue with the tehsildar. At the district level, the land revenue system must work under the Zilla Parishad. The 30 per cent representation for women in PRIs should help the cause of women in so far as recording of rights of women in land is concerned.

Moreover, steps have to be taken to bring about greater transparency in the administration of land records, with access to information regarding land holdings on demand by any individual. Copies of land records, including record of rights, field books/field map, Land Pass Books as also mutation statements, status of land and jamabandi register should be accessible and copies provided on payment of a fee.

LAND TENURE

Relation of the Farmer to the Farm

The use of land may be acquired in two ways: (1) by ownership; (2) by rental. The form of rental varies with localities and systems of farming, but may be either cash or share. Sometimes combination cash and share rental is followed. The question is always before the farmer as to whether it is best for him to own the land or rent it. If he rents the land, which system offers the greatest advantages? These questions cannot be answered to cover all conditions and localities. It has been generally assumed that ownership tended toward the best returns from the land. It is a fact, however, that the best farming in the world probably, is carried on under a rental system. Much of the high priced land of Great Britain and of France is tilled by renters. The farming is intensive and the returns are very high. The advantages to the tillers of the soil in these cases lie in their ability to rent at a low figure. The owners of the land are satisfied to take their returns in "added nobility," which comes from owning land in these countries and from receiving the speculative value accruing in the rise of land. Whether it is best to own or rent, can be determined approximately by careful calculations of the probable crops and products to be grown and comparing these figures with the probable cost of growing them, and with the receipts that may be obtained from the market at hand.

It has been generally assumed by the American farmer that he should own his land. Encouragement has been given to this idea by the fact that through the Homestead Act and others, looking toward the settlement of the land, he has been able to secure possession of the land at low cost. Land values have been rising rapidly and farmers have been anxious to gain the appreciation. Another reason why farmers have preferred to own their land in America lies in the difficulty of getting satisfactory leases for a long term of years. A five or ten year lease would encourage a tenant to grow crops which would nourish the soil and greater attention would be given to keeping out noxious weeds. Fertilizers and manures would be more freely used and the inclination to exploit the soil would be lessened. Short leases result in frequent moves on the part of the tenant, and moving is always attended by more or less loss. Under this system only the poorest farms have been available for renting, and returns have not been sufficient to attract good tenants. Abundance of free or cheap land has tended toward ownership rather than toward developing a satisfactory leasing system.

The Advantages of Ownership

There are several advantages arising from the ownership of land. (1) Ownership assures long tenure or permanent possession. This stimulates an interest in building up the place and making improvements. Land is likely to be better tilled and the fertility of the soil better conserved. Systematic rotation of crops is more likely to be followed, and an effort made to build up a permanent and lasting business. This leads to the erection of good buildings and the maintenance of a permanent home. The owner feels that his interests all lie in the community and he becomes a better citizen from the fact that he owns the land. (2) Ownership gives to the owner the possibility of profits on the appreciation in land value. In this way the occupant of the land gains all of the value resulting from good tillage and the improvement of the land, and he secures the returns from any special forms of improvements that may be added to the property. (3) Ownership of the land obviates vexatious questions of leases and contracts in dealing with landlords. It offers freedom from annoying inspection by the owner, who to protect his property and prevent the depletion of the soil, must insist on certain forms of cultivation, care and cropping.

Against these advantages must be considered the risk incurred, which must be carried by the owner, in a large investment. This risk must be carried whether the crops are good or poor. In poor crop years the money invested in the farm may be made to bear only a very low rate of interest, or the farm may be run at a loss. Buildings depreciate to a certain extent and the owner of the land must bear the depreciation. Land poorly tilled also depreciates in value in certain communities, and if one buys in a poor location and pays too much for the land, he is likely to meet loss from depreciation in land value. The taxes and insurance must be paid and repairs on buildings must be made. These items all add to the annual cost of upkeep and must be counted among the disadvantages of land ownership.

Advantages of Cash Rental

The advantages of cash rental to the renter lie in the reduction of the investment and in shifting the payment of taxes, depreciation and upkeep to the landlord. Where good farms can be rented for cash, this system often brings better results to the renter than ownership, though he loses the possibility of securing the appreciated value of the land. The advantages of cash rental over share rental are as follows: (1) The renter gets all of the advantages of superior tillage and management. If a long lease can be secured, quite as much encouragement is given toward building up good systems of farming as where the land is owned. (2) The renter is not subject to close supervision and inspection by the landlord, which in some cases becomes very objectionable. (3) A greater latitude is allowed on the cropping system and in the management of the farm.

It should be borne in mind, however, that in the case of cash rental, the renter bears all of the risk of crop failure, which in some localities is considerable, due to adverse climatic conditions or to a refractory soil. The advantages of cash rental over share rental to the owner are: (1) Freedom from responsibility in planting, cropping, and in the management of the farm; relief from the task of frequent supervision. (2) It lessens the risk to the owner from crop failure, provided the tenant is responsible. In cases where the tenant is not financially responsible and cannot meet the bills, the owner is forced to lose the rent.

Advantages of Share Rental

Renting land on shares lightens the investment of the renter materially and makes it possible for men to secure farms who have not the capital to buy their farms or rent for cash. In most cases of share rental the renter furnishes the live stock for work purposes, the machinery, and household equipment. The landlord furnishes the other essentials for working the place. The crop is divided on a share basis which lightens the burden of the renter in case of crop failure.

The disadvantage to the renter from share renting lies in the fact that he gets only a share from the increase due to superior cultivation. Often the increase from this cultivation does not pay for the work he is required to do. To illustrate, it may be that a renter will put a dollar's worth of extra work on an acre of land in putting in a wheat crop and increase the yield two bushels.

If wheat should sell at 84 cents, which is the average price for wheat for a ten-year period, and the crop is divided on a half and half basis, the operator would get only 84 per cent of the dollar's worth of work. The fact that the increase must be shared with the landlord and that there is a probability of getting only slightly larger returns on good tillage, results in share-rented farms being poorly tilled and unsystematically cropped.

Ownership, Cash or Share Rental

It is possible to calculate approximately the returns from the land under each system of land tenure, as the following figures will illustrate. On a farm of 160 acres in the Central West where farm records were made the figures under the ownership system were as follows:

In the land	...	$12,000.00
In buildings	...	4,000.00
In live stock	...	1,162.00
In tools and machinery	...	389.00
Making a total investment of	...	$17,551.00

This amount should bear at least 5 per cent interest. The expense of operating the farm under the ownership system was,

Interest on investment	...	$877.55
Building repairs and depreciation	...	160.00
Machinery purchased	...	331.00
Taxes	...	85.00
Insurance	...	15.00
Seed	...	80.00
Stock purchased	...	256.00
Feed and supplies purchased	...	62.00
Labour	...	33.00
Cash for incidental expenses	...	50.00
Total	...	$1,949.55

The receipts from the farm were,

From crops	...	$1,278.00
From live stock products	...	105.00
From sales of live stock	...	290.00
Increase in inventory	...	800.00
Total	...	$2,473.00

This would leave a net income for labour of $523.45, after 5 per cent interest has been paid on the investment. Under the cash rental system the interest on the land, depreciation on buildings, taxes, and insurance would be shifted to the owner. The operator would have the following expense:

Interest on money invested in live stock,		
tools and machinery, at 5%	...	$77.55
Cash rent, 160 acres at $4.25	...	680.00
Machinery purchased	...	331.00
Personal taxes and insurance	...	15.00
Seed	...	80.00
Labour	...	33.00
Feed and supplies	...	62.00
Stock purchased	...	256.00
Cash for operating	...	50.00
Total	...	$1,584.55
The receipts would be as before,		
From crops	...	$1,278.00
From live stock products	...	105.00
From live stock sales	...	290.00
Increased inventory	...	800.00
Total	...	$2,473.00

This leaves a net balance or labour income to the renter of $888.45, a gain of $365.00 over the ownership system. It should be borne in mind, however, that this gain is offset by the fact that the renter does not gain the advantage of rise in land value. The land owner receives only $680.00 for the use of his farm which is only 2.5 per cent on its net value. It is the possible appreciation in land value that satisfies the owner with the low interest rate received. He is also free from the care and labour of managing the farm. Under the share-renting system on a lease where the tenant owns all the live stock, feeding them with his own feed, and where the crops and hay are equally divided and the landlord furnishes the seed, the results to the renter would be as follows:

Expense of Tenant

Interest on money invested in live		
stock, tools, and machinery at 5%	...	$77.55
Machinery purchased	...	331.00
Personal taxes and insurance	...	15.00

Labour	...	33.00
Feed and supplies	...	62.00
Stock purchased	...	256.00
Cash for operating	...	50.00
Total	...	$824.55
Receipts of Tenant		
One-half of crop sales	...	$639.00
Live stock products sold	...	105.00
Live stock sold	...	290.00
Increased inventory	...	593.00
Total	...	$1,627.00
Expense	...	824.55
Receipts less expense	...	802.45

The tenant would receive a labour income of $802.45 which is $86.00 less than he would have received under the cash renting system, but $279.00 more than he would have made in owning the farm himself. The reward for the loss of the $86.00 is in the freedom from responsibility for a large cash indebtedness and lessened risk from poor crops. The immediate returns from share renting would be greater than from owning the farm, but the one who rents sacrifices the possibility of gain from rises in land values. In localities where depreciated values are likely to be met, a share rental system would be preferable for the farm operator. The land owner would receive for his share $639.00 from sale of crops, less $80 for seed, or $559.00. He would also have one-half of the hay. To make as much under the share-renting system as under the cash renting, he would have to sell $121.00 worth of hay.

FARM EQUIPMENT

The equipment of a farm is a matter that has a close relation to the possible profits from operating it. The farmer must have sufficient work stock and machinery for operating the place easily and well. He must also have suitable buildings for storing his produce. Chapter III suggests a proportion for the investment of capital which will give satisfactory results in the general organization of the farm. The individual farmer, however, must study the problems which confront him, and erect buildings or buy machinery only as it can be shown that they will add to the profit form his farm, or lessen the labour of himself or the family, or add to the comforts and pleasures of the home. Some farmers will need corn cultivators, others need potato diggers, while still others need grain binders, hay loaders, or manure spreaders. Even in the same community, there is a great difference in the needs of the individual farmers for equipment or buildings.

Buildings, an Expense to the Farm

Few buildings return a direct profit on the money invested in them. It is necessary for the farmer to have a house for his family to live in, but it really does not add to the production from the farm. It is rather a bill of expense. The depreciation and upkeep will amount to five per cent or more on the investment.

So far as securing profits from the operation of the farm is concerned, therefore, it is to the farmer's interest to invest as little as possible in the house. A good house undoubtedly adds to the value of a farm as a home and increases its selling value, but does not in any way increase the production from the farm. The desire for comfort and the personal satisfaction of living in a good home must be weighed against the possibilities of paying the bills from the earnings of the farm.

The average sized farm in the United States is 138.1 acres. In Minnesota it is 177 acres. Few of the Minnesota farms under the present system of farming can afford to pay interest on an investment in a house for the farmer and his family of more than $2,000 to $2,500.

If a house costing more than that amount is erected the additional investment should be charged as a personal expense to the individual, and not as an operating expense against the farm. On many small farms a house costing $1,000 to $1,200 would adequately care for the labour necessary to operate the farm. In other states and particularly in the South, the dwellings will cost less than the amount indicated. The practice should be to charge against the operating expense of the farm only such dwellings as are used to shelter the working force of the farm, or in some way add to the farm income. In cases where a tenant house is provided to house and feed the farm labour, the owner's house should not be charged to the operating expense of the farm, but as a personal expense to the farmer.

Barns and other building likewise, must be made to add to the income of the farm by sheltering productive live stock; by protecting feed or market products from deterioration; or by affording shelter to productive labour. If they do not, they become a bill of expense to the farm. There should be a well demonstrated need for a building before its erection is undertaken.

What Machinery to Buy

The farmer is often confronted with the question of what machinery to buy, what size would be most economical and how to care for it so as to get the greatest use from it and make it last. As a rule, the American farmer is inclined to overstock with machinery. New inventions have been numerous and manufacturers are good advertisers. As a result, the farmers are led to try many machines before their value has been established or their wearing qualities determined.

The consequence is, that many must soon be discarded as poorly adapted to the work to be done. It is best not to be too hasty in buying new inventions. Experiment Stations and wealthy farm owners should do the experimenting. Farmers who have to pay for their machinery from the earnings of their farm, should stick to the "standard" machines.

Calculate when to Buy Machinery

Investments in machinery should be made only after carefully calculating the probabilities of being able to make the machine pay interest on the investment. This can be done by considering, (1) The quality of the work that can be done with and without the machine. (2) The cost of the work, with and without. (The cost includes interest, depreciation, cost of care, storage, insurance, risk, and repairs.) (3) The returns form the same capital invested in live stock, labour, or other enterprise. (4) The gain or loss in quality of product, with and without the machine. (5) The saving of man or horse labour. Other factors such as greater amount of time for recreation, or for other more agreeable work, should also be considered, though they do not necessarily affect the profits.

The original or first cost of the machine under consideration, should not be made the sole factor in the question. The cost per acre or per unit of product, should be made the basis of machinery cost and investment. The total cost for a year including interest on investment, repairs, depreciation, oil, and shelter, divided by the number of acres handled, will give the cost per acre for the machine. Calculations of the cost per acre of doing the same work without the machine should be made also and a method reached that will give the greatest possible gain.

Size of Machinery to Buy

Machines are made in various sizes. In buying a new machine, a farmer should consider at least two things. First, the amount of work to be done; second, the motive power available. It is a mistake to buy too small a machine and just as great a mistake to buy a machine so large that the motive power cannot adequately propel the machine. In determining what size to buy the capacity must be considered. The following example will illustrate. A farmer has ten cows, each giving 5,000 pounds of milk yearly. The aggregate amount of milk will be 50,000 pounds a year. Which will be most economical to buy for separating the milk, a separator with a capacity of 450 pounds per hour costing $100, or one with a capacity of 600 pounds per hour, for $125? In separating 50,000 pounds of milk with the 450 pound separator, 111 hours will be required, which at 14 cents per hour, will cost $15.54. With the 600-pound machine, 83 hours will be required costing $11.62. there will be again in labour saved, therefore, of $3.92. This, however, is not all profit. There is an additional investment in the machine of $25.00, which at 6 per cent would bear $1.50 interest. The depreciation also must be considered on the additional cost of investment. It is estimated that the depreciation on milk separators is 5 per cent. Five per cent on $25.00 will be $1.25. This added to the interest on the additional investment, amounts to $2.75. Subtracting the additional interest and depreciation charge of $2.75 from $3.92 saved in labour, would leave a net gain of $1.17 per year in the use of the larger machine. Similar calculations can be made comparing the gain to be made from the purchase of a large or small drill; the purchase of small walking plows or a large gang plow; the use of large and small grain binders; and many other machines used about the farm.

Maintenance

There is a great difference in the rate at which different farmers wear out machinery and equipment. The good care-taker can do much to prolong the life of his machinery. Proper housing is essential to keeping machinery in good order, but good housing is expensive and adds to the machinery cost of the farm. Machines that stand outdoors, warp and twist badly, thus increasing the friction when they are put to use. Machines stored in poor machine sheds with leaky roofs and wet floors, often rust quite as badly as if they were standing outdoors. Attention should be given also to the care of machines while in the field. Few operators take the time and care to properly oil the bearings. bolts are allowed to become loose, thus letting the machine sag or get out of adjustment. The matter of adjustment of the bearings and cutting devices also, is a factor in determining the length of time the machine can be used. Attention to these small details will add very much to the usefulness and length of life of a machine.

It is a good practice to go over the farm machinery at least once each year, tightening up bolts, renewing worn portions and making sure that the bearings are in good order for operation. If a forge and some blacksmith or carpenter tools can be had, much of the ordinary repair work can be done by the farmer or his sons. The saving in the cost of the work will not be great perhaps, but the saving in time, in going to the shop in the village, may amount to considerable.

Much of this repair work can be done on rainy days or in winter when farm work is not pressing, and the machines can be put in good condition then for the busy days during the seeding and harvest seasons. The depreciation of farm machinery is illustrated which is taken from Bulletin No. 117 of the Minnesota Experiment Station. These figures cover the depreciation of machinery on a number of Minnesota farms where records were kept during the years from 1902 to 1907.

A large variation in the rate of depreciation on various machines. Manure spreaders, hay loaders, and threshing outfits depreciate quite rapidly in value. Hay tedders, fanning mills, and grain tanks or wagon boxes, depreciate comparatively slowly. The rate of depreciation will not be the same on all farms, but by inventorying the cost price, and making an arbitrary valuation of the machine each year, it will be possible by using the figures covering ten to fifteen years, to determine the rate of depreciation on any farm. Most farmers will not be in position to do this. Where they do not apply, it will probably be safe to estimate the depreciation on machinery at 10 per cent.

Machinery Cost per Acre

It is a matter of interest to know the cost per acre for the various kinds of machinery used on the farm. It may be made a factor in determing [*sic*] the cost of producing crops. The cost per acre is determined by including all charges against each of the machines for the year, such as interest, depreciation, repairs, oil, and cost of housing. This cost for each machine divided by the number of acres handled, determines the cost per acre for the use of the machinery.

Duty of Machinery

The capacity of machinery is often determined by the motive power applied. A machine cannot be run at full capacity unless sufficient motive power is supplied. The efficiency or capacity of a machine is based on horse power or the speed at which it is driven. Few machines perform all of the labour of which they are capable, owing to the fact that the use of the machine is interrupted by the demand for the farmer's time in doing chores which keeps him from the field work. In other words, the machines are not used for the full length of time for which they are capable.

RELATION OF LIVE STOCK TO FARM MANAGEMENT

Live Stock and Soil Fertility

The opinion commonly prevails that live stock raising helps to maintain the fertility of the soil. This is true only under certain conditions. An oats crop yielding 50 bushels per acre will require from the soil 50 pounds of nitrogen, 18 pounds of phosphoric acid, 45 pounds of potash, besides small amounts of numerous other elements.* If the grain is sold and the straw is burned, the entire amount may truly be said to be removed form the soil.

If the grain is fed to cows, the milk or butter-fat sold, and the straw used for bedding, but the manure is allowed to remain in the yard or barn, just as much fertility is removed form the land as though the crop was sold and the straw burned. However, if in feeding the grain and using the straw for bedding, care is taken to save all of the manure an dlitter, returning it to the fields from which the crop came, there will be little loss of fertility.

It is very difficult to prevent some loss through decomposition, evaporation, or leeching. There may also be much or little loss from mechanical sources. This will depend upon the methods employed in getting the manure to the field. The loss from all of these sources varies from forty to sixty per cent. About three fourths of the fertility contained in the feed is recovered in the manure as it comes from the animals. This may all be returned to the field where leeching and mechanical losses are prevented. On most farms, nearly half of the fertilizing value of the feed is lost before the manure is returned to the field.

The animals through which the foods pass add nothing to the fertilizing qualities of the manure. The food stuffs if returned directly to the field upon which they grow, would quite as well maintain the fertility of the soil. In case of the growth of certain of the crops, as clover and alfalfa, the fertility may even be increased through the addition of nitrogen which these plants have the power to gather from the air in the soil.

The mastication and digestion of the food stuffs by animals, however, renders available more quickly the fertilizing elements of the plants. It is this factor which is largely accountable for the belief that the process of mastication and digestion has added to the fertilizing value of the food stuffs. Animal manures without question are one of the best possible sources for maintaining soil fertility, but the farmer should recognize the fact that to get their full value, he must prevent all mechanical and chemical losses and that they must be distributed evenly over the farm and incorporated with the soil.

How to Save the Manure

So far as possible, the best way of saving the manure and preventing loss of fertilizing elements, is to haul it from the barns as made, directly to the fields, spreading it at once and incorporating it with the soil. Besides saving the fertilizing value of the manure, this method of application has the added advantage of economizing labour as well. It requires no more time to throw the manure directly into a manure spreader or wagon, than to wheel it out of the door and throw it in a pile, which must later be loaded and hauled to the field. Less labour is required than where the manure is loaded into a wagon and hauled to a compost heap from which it must again be loaded into a wagon or manure spreader and taken to the field.

It is not possible on many farms to haul manure from the barn directly to the fold throughout the year. Growing crops frequently interfere during the summer season at least, and the other work of the farm often prevents. In the North, snow and hard freezing weather make it difficult to operate the manure spreader in the winter time.

The manure can be hauled on a wagon or sleigh and spread by hand, but it is at the expense of uniformity of application. Manure so applied should be spread after the frost leaves in the spring. This may be don by spreading out the thick places with a fork or by going over the field with a spring tooth or smoothing harrow. Whenever land is available, however, upon which to spread the manure, it should be taken directly from the barns or yards to the fields and spread evenly and ratherthinly as hauled. In many places the practice prevails of hauling manure to the fields and unloading in half load to load piles.

It is allowed to stand in these piles during the winter and is then spread by hand in the spring. Nearly as much loss occurs from decomposition and leaching where manure is handled in this manner, as where it is allowed to lie in piles under the eaves of the barn. The labour of applying is also increased and the fertilizing value is not uniformly distributed. Where manure cannot be hauled directly to the fields, it should be spread out evenly over a level yard and the stock allowed to run over it, tramping it firmly. If the yard can have a concrete bottom there will be less loss from leaching than where it lies on open ground. In the East where the barn-yard manures are highly valued, manure pits are made and frequently covered with a roof. If the manure from all classes of animals can be mixed together in such a yard, or pit the loss from heating and evaporation will be very much lessened. In countries where the population is very dense and all of the manure must be saved, cisterns are made in which the liquid excrement may be collected. The liquid is pumped into tanks and hauled to the fields. Cement pits are also constructed in which both solid and liquid excrement are collected for decomposition. Such systems avoid the loss of fertilizing elements but are epensive.

Live Stock Consumes Unsalable Products

On most farms, large quantities of cheap rough foods are grown for which there is no market or which cannot be transported to market profitably. Straw from the various grains, corn stover, corn fodder, ensilage, and grass aftermath are examples of this class of food stuffs. Many farms contain a portion of rough land which is difficult to till. This land often makes desirable pasture and can be made to return something of profit to the farm by the use of live stock. The cheapest foods are those on which little labour is expended in growing. The cheapest way to feed foods is to let the animals gather them themselves. Pasture, even on comparatively expensive land, is one of the cheapest animal foods that can be provided. It is the natural food of most classes of live stock and naturally adapted to their needs. As a consequence, they gain rapidly on it or yield a large amount of product, if protected from annoying storms and insects. It rquires but little labour to provide pastures for live stock. When a good sod has been formed all that is necessary is to maintain the fences, keep out the weeds, and dress up the thin spots. The amount of labour is much less than would be required to grow a crop and harvest it for hay or forage. The silo fills an important place in providing cheap forage for live stock. Corn and other bulky, heavy yielding crops are preserved in very palatable and digestible form without waste by running them through an ensilage cutter and storing them in a silo. Silage is especially desirable for feeding dairy cows. It is also a valuable food for beef cattle and sheep. It is by making use of cheap forage and pasture that the farmer can hope to support large quantities of live stock, which by being stabled and yarded in the winter, will enable him to provide manure for the land that is not pastured.

Profits come from Cheap Crops

Investigations of cost of production indicate that the greatest profit from live stock is made where the animals are fed on these cheap foods. Where chickens, pigs, or cattle are fed on highly concentrated marketable products, it has been possible to make but very littleprofit. Where chickens are confined in close yards they have frequently been handled at a loss. On farms where they are allowed to run in the barnyards and through the fields, picking the larger part of their living, a flock of fifty to one hundred chickens appear to cost nothing for feed. This is because they exist on natural foods that would otherwise be wasted. Likewise, the farmers who have provided good grass or annual pasture for their brood sows and pigs, or who have allowed them to run in the fields gathering their own forage during the summer months, and who have finished them by turning them into a corn field allowing them to pull down and harvest their own feed, have made good profits at hog raising. Those who have confined their hogs in close yards, have expended labour in mixing feed and carrying it to them, and in husking corn and feeding it to them in small dry lots, have seldom made large profits.

It is commonly believed that dairying is a profitable business. This opinion is due to the fact that the farmers who have a few cows seem to have better crops and appear to have more money than those who do not raise cows. An analysis of the business shows, however, that the main income from the farm in many cases is not made from the cows, but comes from some grain crop or other product. The cows do add to the profits of the farm by utilizing the cheap feeds and converting them into a concentrated salable market product. Most of the dairy product of the United States is made by low grade cows in which little money is invested, feeding on coarse feeds from which no returns would be made otherwise. It is the conversion of this material into a market product which makes the cows popular and profitable to the average farmer.

On a farm of 160 acres, four horses ordinarily would be required to do the work. The time of these horses would be so fully occupied that it would not be wise to undertake to raise colts from any of them. By adding another horse it would be possible to raise colts from two of them and still do the work of the farm. The work mare will do as much work as a gelding if she does not raise a colt. She will do nearly as much when required to raise a colt, but will not last so long if required to work full time. For this reason it is good policy to allow some leisure time just previous to and shortly after foaling. Colts raised by farm work mares offer a reasonable source of profit to most farmers.

In planning for the crops to be used as food for live stock, it is well to consider not only yields but feeding values as well. An acre of corn will yield 12 tons of ensilage which will contain approximately 3600 pounds of digestible nutrients. An acre of timothy yielding 2 tons would contain only 1700 pounds of digestible nutrients, of which about 110 pounds is protein. An acre of clover yielding 2 1/2 tons of hay would contain 2300 or more pounds of digestible nutrients; 350 pounds of this would be protein. Alfalfa yielding 4 tons per acre, would give 4100 or 4200 pounds of digestible nutrients, of which 840 pounds or more will be protein.

These nutrients will not all be produced at the same cost, however. An acre of hay of any kind can be produced at a much lower cost than an acre of ensilage. The labour cost of providing the foods must be considered as well as the value of the nutrients yielded by the various crops. It costs about $5.60 to produce an acre of timothy hay and $20.60 to produce and store an acre of ensilage. The food nutrients in hay would, therefore, cost only 3.29 cents a pound, while the nutrient in the ensilage would cost 5.72 cents. The succulence of the ensilage gives it a greater feeding value than is shown by the analysis, however, and it will be found a most profitable food for the classes of animals called ruminants.

It should be borne in mind that the one who feeds salable grain, such as wheat, corn, and oats, to live stock, is in competition with the farmers who raise their stock on grass, corn fodder, ensilage, and other cheap waste products of the farm. To compete successfully with them, the quality oproduct must be higher or more must be produced for the same value in feed. For this reason, those who feed large numbers of animals on high priced grain, usually raise pure bred stock which has a breeding value as well as a market value for product produced. In this way a margin of profit may be secured above the cost of the feed.

Live Stock Provides Employment for Cheap Labour

On most farms there are children old enough to perform some light labour. They can often be profitably employed in caring for the poultry, sheep or other live stock. In many cases most of the milking is done by the farmer's wife and children. They are not paid wages for doing the work and the product secured from the expenditure of their labour is nearly clear gain, as they must be fed and clothed whether they work or not.

Likewise, where two or three cows apiece are milked by the hired man after a full day's work in the field, the product is secured at a low cost so far as labour is concerned. It is because live stock raisers are thus able to give employment to otherwise nonproductive labour or to labour that is not fully employed, that they often make larger farm incomes than those who raise hay or grain only. A small amount of live stock can be cared for without appreciably reducing the amount of labour available.

Live stock raisers who have children old enough to work, are thus placed at an advantage over live stock raisers who must hire all labour used. While farmers do not always realize the facts in the case, it is true that most of the profits from live stock raising are derived from these two sources: (1) A conversion into concentrated marketable products of cheap bulky foods that would otherwise be wasted; (2) By the use of labour that does not have full value but which must be supported, whether employed or not. The margin of profit is seldom large but the income on a stock and grain farm is less likely to fluctuate than where grain raising only is followed.

Kind of Live Stock to Keep

The kind of live stock to keep depends upon several factors. It has not yet been proven that one kind of live stock is more profitable than another. Some men make excellent profits in raising horses while others fail. There are many farmers who make generous labour incomes from keeping dairy cows, but the average cow barely pays for the feed consumed and the interest on the investment and the labour cost of caring for her. A few farmers have become rich from raising beef cattle, or raising hogs or sheep, but in all of these cases it is the man and methods of management, not the particular kind of stock that are responsible for the profit.

The tastes of the individual farmer are an important factor in the decision. A man who does not like cows and who "hates" to milk will do better in some other line of stock raising than dairying, though he may learn to love the cow and overcome his dislike for milking if he finds a good profit in doing it. The man who "loves" a horse and dislikes a pig is more likely to make a profit on horse raising than on pig feeding.

The size of the farm must also be considered. As a rule, small animals go with small farms. A twenty-acre farm may be well adapted to poultry raising and would do fairly well for a small number of pigs or sheep, if well farmed. It would not be at all adapted to dairying, stock feeding, or horse raising, as too large a proportion of the feed would need to be bought. Pasutre and rough feed could not be supplied and labour in caring for the stock would be expensive on account of close confinement. While a few horses or cows might be kept, it would be impossible to build up a big business on so small a farm.

Surveys of typical farming areas indicate that it commonly requires three to five acres of good arable land to support a horse. Two and on-half to three acres will be required to support a cow. Three-fourths of an acre to one and one-fourth acres for a hog, and one-fourth to one-half acre to support a sheep. Consequently, on a farm of 160 acres, 32 to 53 horses would be the maximum number that could be kept if all other live stock was excluded and all of the land devoted to growing horses. Few farms are so intensively stocked. A combination of two or more kinds of stock will usually be made.

Market demands and shipping facilities must also be considered in deciding what kind of stock to keep on a certain farm. Live stock products are concentrated and easily shipped if transportation is rapid and frequent. Milk and cream are perishable, however, and must be marketed daily or at least every other day. Eggs also may deteriorate in reaching market if shipping facilities aroor. Live animals may be shipped long distances but are subject to heavy shrinkage or loss by accident or death. Freight charges and care while on the road must also be covered. Short distances are most favorable for live stock shipment. Grain and other imperishable products may more safely be shipped long distances.

The amount of capital to invest often determines the kind of stock that should be kept. Usually a farman be more cheaply stocke with the small animals than with the large ones. There is also need of considering, in connection with the investment, the length of time which must elapse before returns will come in. In this respect the advantage lies with the cows, as the milk and butter-fat are immediately available for sale. With sheep, the lamb or wool crop is likely to mature within six months. The returns from hogs can rarely be realized short of six or eight months, while in horse raising, the colt crop is not mature for market inside of three years, though younger colts are often sold. Better prices can be secured, however, when they are over, rather than under, three years of age.

Limitations in Stock Raising

There are certain limitations to each kind of stock raising which should be considered also. The number of colts raised should be determined by the number of mares that can be profitably employed on the farm. Horses must be raised as an adjunct to other lines; unless they are, the first cost of the colts is too great. It costs from $60 to $100 per year to keep a mare and unless she performs some labour, the colt will not be valuable enough when a year old to meet the charge for keeping the mare. On the range or on very cheap land, the cost may be reduced and colts profitably raised, even though the mares are not worked. On the arable farms, the first cost of a colt including stallion fees, can be reduced to $50 if the mare is required to perform a reasonable amount of work.

The number of dairy cows that can be kept is limited by the amount of pasture that can be furnished or of roughage grown. It is not profitable to buy or ship roughage. Milkmen located near favourable markets may be able to produce milk at a profit, even though the rough feed must be shipped, but when applied to the business as a whole, it cannot profitably be done. Another factor which limits the number of dairy cows that can be kept is labour. One man cannot feed and cere for more than ten to twelve cows, as not more than that number can be milked in the time usually available for milking. Some other farm work can always be profitably combined with dariryng. It is because the labour is better distributed, that the farmer who grows some crops and keeps some cows, makes a better profit than the one who conducts either line alone.

Beef cattle raising is limited by the amount of grazing land. The business cannot be conducted on small areas. Cheap pasture and forage are essential to secure profits. Beef cattle raising must not be confused with stock feeding which can profitably be carried on, on small areas of land, provided the proper feed can be purchased. One man can care for 50 to 60 head of beef cattle. Large numbers must be kept together to reduce the cost of care. They are best adapted to large farms or to the ranges.

Sheep raising cannot be conducted on low wet land. Foot-rot and other diseases are sure to attack them on such land. They require high land and dry quarters. Pasture and cheap roughage are essential. They do not do well in close confinement except for short periods. If kept in close quarters they are likely to be infested with vermin and to become infected with parasic diseases. The best profits will be made on a few sheep kept as scavengers on a diversified farm.

It has been customary in the past to keep swine in close quarters. They require large quantitites of concentrates, but can use profitably good clover, alfalfa, or rape pasture. These crops are especially valuable for raising small pigs. The number that can be kept is determined by the amount of concentrated food that can be grown and the amount of pasture furnished. Swine are subject to cholera, swine plague, and other diseases, and should not be herded in large numbers.

Pure Bred vs. Grade Stock

Pure bred stock is often advised. For productive purposes it is doubtful whether pedigreed animals are any better than those without pedigree. Grades of the right conformation and descending from families of heavy producers will yield as great a profit for the market product as pure bred stock. The great advantage of pure bred stock is the guaranty that they do come from good families. This gives them a breeding value which can often be sold to advantage.

To be worth more than grade stock, they must come from highly productive families and attention should be given to keep them up to a high standard. When one can combine good breeding value with large production of market commodities at a low cost, the profits can be greatly increased. It is for this reason that pure bred stock raising can be made to pay large profits. If market products alone are desired, grades may be kept quite profitably.

When is a Farm Well Stocked

It is difficult to tell when a farm is fully stocked, or to know whether one farm is carrying as much live stock as another one. It is especially difficult to compare a farm on which horse raising is the main line of live stock, with one on which hogs and cattle may be kept. G. F. Warren, of Cornell University, New York, offers a basis for comparing the amount of livestock kept. This comparison is made on the basis of feed required and value of manures produced. The figures will vary somewhat as one farmer may feed more heavily than another, but it is believed that the figures offered are fairly exact and that satisfactory comparisons can be made. For the purpose of comparison, one cow, bull, steer or horse, two years old or older, is called an animal unit. Two head of colts or young cattle are counted as a unit. Seven sheep, fourteen lambs, five hogs, ten pigs, and 100 hens may each be called an animal unit. On a farm having six horses, two colts, twelve cows, eight young stock, six brood sows, thirty pigs, twenty sheep, and 100 chickens, there would be approximately thirty-two animal units. On a 160-acre farm, this would be one animal unit for each five acres of land. This is considered a fair proportion. By counting the number of animals on other farms and reducing them to animal units, a comparison may be made to determine which farm is stocked more heavily. The knowledge is valuable also in determining how much manure can be supplied and how often an application can be made to the fields.

FARM NUMBERS, FARM TYPES, AND CONSERVATION PROGRAMS

The number of farms has declined dramatically since its peak of 6.8 million in 1935, with most of the decline occurring during the 1940s, 1950s, and 1960s. The decline in farm numbers has levelled off since the 1970s. By 2002, 2.1 million farms remained. The remaining farms have a much larger average acreage, but averages mask differences among farms. Today's farms range from very small retirement and residential farms to industrialized operations with sales in the millions. Part of this diversity stems from the very low sales threshold ($1,000) necessary for an operation to qualify as a farm for statistical purposes.

One way to address the diversity of farms is to categorize them into more homogeneous groups. The farm typology developed by ERS identifies five groups of small family farms (sales less than $250,000): limited-resource, retirement, residential/lifestyle, farming-occupation/low-sales, and farming-occupation/high-sales. The typology also includes large family farms, very large family farms, and nonfamily farms. In addition, very small farms (sales less than $10,000) make up more than half of all farms. Very small farms account for a particularly large share of farms in the limited-resource (72 percent), retirement (76 percent), and residential/lifestyle (76 percent) groups. Production, however, is concentrated among larger farms; small farms account for only 27 percent of the total value of production.

The smallness of most farms has implications for conservation and the environment. An ERS study found that smaller corn farms are less likely to use conservation tillage than are larger farms. The practice is more practical for larger farms because they have more acres over which to spread the cost of new or retrofitted equipment necessary to adopt conservation tillage. Small farms whose operators are retired or farm part-time are also less likely to adopt conservation tillage, possibly because of hesitancy to change familiar production practices. Small farms, however, participate widely in the Conservation Reserve Program (CRP) and the Wetlands Reserve Program (WRP).

LONG-TERM TRENDS AND AGRICULTURAL POLICIES

The Partition, both in India and in Pakistan. Total output growth rates turned from zero or little to significantly positive levels, which were sustained throughout the post-independence period; crop mix changed with increasing concentration since the mid 1950s.

Therefore, as far as the data suggest, agriculture in both countries has been experiencing one-way concentration of crops since the mid 1950s, when agricultural transformation in terms of output per agricultural worker was proceeding.

According to Timmer's (1997) stylization, this is a stage before a mature market economy with diversified production and consumption at the national level. Against our initial expectation, these trends were not reversed in the 1990s in the direction of more diversified production if we focus on major crops.

The trends were strengthened in India during the 1990s while they became less significant in Pakistan. The performance in the 1990s suggests that agriculture in India has remained in the phase of diversification where concentration into high value-added crops is strengthened.

To investigate the impact of institutional and policy changes in the 1990s, we compare major changes in political institutions compiled with long-term trends in production. The colonial period in the early twentieth century was associated with free trade of agricultural commodities. Under this institutional setting, public investment into irrigation contributed to the rapid expansion of production, especially in regions currently belonging to Pakistan. The colonial period ended with increasing state control on trade and marketing under the war pressure. This period was associated with stagnation in agricultural production.

THE BUSINESS SIDE OF FARMING

Farming as a Business

Farming has not usually been considered a business. The diversity of the duties of the farmer, the area over which the operations of the farm extend and the complexity of the records required, combine in making difficult the organization of the details of farming into business form. The successful financial operation of a farm presents quite as complex problems and calls for at least as much business ability and judgment as is required in operating a store with the same investment.

Farming, therefore, should be considered as a business, and the man who can produce his crops and products at the lowest cost and sell them at the highest price, investing the proceeds to the best advantage, should be considered the best farm manager. The man who knows the details of the cost of production and operation, and whose records show the profitable and unprofitable lines of production, thus enabling him to eliminate those that do not yield a profit, may be counted as the best business man.

A farmer should know the elements of soil fertility. He must understand the principles of the movement of soil water, and the action of soil bacteria. He should understand the nature of plant growth and be familiear with varieties and species of plants and with the effect of one crop on the crop following. He must also be familiar with animals and their habits and know how to feed and care for them. In addition, he must know how to buy and sell to advantage, make contracts, and plan his buildings and his farm so as to necessitate a light expenditure for labour, also that he may distribute his labour to advantage over the various farm enterprises. And he should know how to keep accounts. The farmer in organizing his business could well follow the example of the merchant. The merchant first takes an inventory of his stock. He studies the demand for his goods, both present and prospective. He notes the supply, the cost, and the demand for each article.

He calculates the labour required to operate his business and such other items of expense are considered as may be legitimately charged against the business. He regulates his purchases and his prices according to the cost of securing his goods and putting them on the market. In conducting a large store business, it is customary to organize it into departments, putting some competent person in charge of each department and having the labour and accounting charges so systematized and recorded as to show the profit or loss from each department and from the business as a whole.

The farmer should likewise take an inventory of his capital, stock, and equipment. He should consider the type of farming to which the soil and climate are adapted. He should consider the fertility of the soil and the demand that will be made upon it by the crops grown.

He should consider, in connection with the soil fertility, the sources from which it may be renewed and at what cost. He must study the markets, the transportation, and the demand for such crops as he grows; also the cost of producing each of the crops and the probable net profit that will be returned. His labour likewise should be charged against the various crops or enterprises and distributed to the best advantage.

In studying the problems of farm organizations, interest on investment, taxes, insurance, and other expense must be included as they affect the financial result. As in a large store business, it is frequently necessary to organize the large farm into departments, keeping accounts with the dairy, with the swine, the grain crops, the garden, and other similar enterprises. Where the business is large enough, it is well to put an expert in charge of each large branch or group of enterprises, thus enabling one to use cheaper labour for performing the work or making the labour more effective by closer supervision.

Where the farming is conducted as an organized business, and accounts are kept with the various lines of work, it is possible at the end of the year, to make a business statement which will show which lines have been profitable. The manager then can change his methods or drop out those lines that prove to be unprofitable and the business as a whole may be put on a better basis.

Investment

The investment of money in land, buildings, and equipment demands careful consideration. It is possible to pay so much for a farm that it will be impossible to produce sufficient revenue to meet the expense of operation and to pay a normal rate of interest on the money invested. This is particularly true where low-priced products are produced. A farm may be highly productive but so located that it will be impossible to market the produce on a profit bearing basis. One should study closely the market facilities of the neighbourhood and raise supplies which can be successfully marketed locally, or which can be transported to a market that pays well for such produce. Unless the produce of the farm is well related to the market, the farm is likely to be operated at a loss.

Proportion in Real Estate

It is a mistake for one to invest all of his capital in the real estate itself. Sufficient capital should be reserved for operating the farm. The hunger for land has induced many farmers to buy more land that they can equip and operate well. Such farmers are said to be land poor. They would secure greater profit from a medium sized farm, well tilled and managed, than from a large one which is insufficiently equipped and poorly operated. Rarely should more than 50 or 60 per cent of the capital be tied up in the land. The size of the farm and the amount of equipment are closely related to the possible profits. A farm of forty or eighty acres devoted to diversified crops and live stock, cannot be so economically equipped per acre as a larger farm. The investment per acre in machinery will be higher as the cost will be spread over fewer acres than the machinery has capacity to handle. Investment in other equipment will also be correspondingly high. Often the labour on such a farm is not fully employed and loss results from inactivity of labour and equipment. A medium to large sized farm, when well organized, fully equipped, and with sufficient capital reserved to operate it well, will pay a much better labour income than a small farm.

Proportion of Investment in Machinery

Investments in machinery are worthy of quite as much consideration as investments in land. Machinery is looked upon as one of the means of reducing the cost of production. The wise use of machinery saves time and labour and enables the farmer to handle large acreages. In this light, the use of ample machinery is wise. The fact remains, however, that investments in machinery are often poorly made and that many farmers are embarrassed by debts for machinery. Frequently, farmers purchase machinery because it is fashionable or because a neighbor has it, rather than because carefully made calculations show that a certain machine can be used profitably.

A safe rule is to buy no machine until carefully made calculations show that the cost of production of a certain crop or product will be reduced sufficiently by the purchase to cover the cost of the machine. That machinery investments can be studied from a business standpoint is quite plain. The following example, showing the relative cost of cutting corn with a machine versus cutting by hand, will illustrate:

The original cost of a corn binder is $125. The annual depreciation, as shown by statistical records covering ten years' work with farmers in Minnesota, is $12.50. The interest on the investment at the average value of the binder throughout its life will be $4.12. Repairs, shelter, and insurance will cost $2 annually. The total annual cost for the use of the binder, therefore, will be $18.62. If only twenty acres of corn are grown each year, the annual cost an acre for the corn binder will be $.93. The cost of cutting corn would be as follows:

It will be noted that a twenty-acre corn field can be more economically harvested by hand where labour is available. On a ten-acre field, the difference in expense would be still greater in favour of the hand harvesting, as the cost of machinery per acre would be doubled. The scarcity of labour, however, and the necessity of harvesting corn quickly to save it from frost, would often warrant the expenditure for the machine, even though it does slightly raise the cost of harvesting per acre. The greater the acreage, the more useful the machine becomes in harvesting, and the less the expense per acre for machinery use.

Calculations similar to this should be made before purchasing a machine for any purpose. If it can be shown that the cost of performing the labour may be reduced by the machine, and that the labour can be performed in a more satisfactory and efficient manner with the use of it, then the purchase may be warranted. In many cases, however, calculation will show that the purchase is not warranted and that it would be better to hire labour and rent the machine, or to buy in partnership with some one else.

Cost of Motive Power

Another factor that should receive consideration is the cost of motive power in use on the farm. Horses usually furnish the farm motive power, though tractors are used to advantage in some cases. Auto trucks and automobiles can often be used to advantage in marketing dairy, fruit and garden products and the cost of using them, when it can be ascertained, should be compared with the cost of horse power. It costs from $40 to $100 per year to keep a work horse, depending on the locality and on the price of feed stuffs; also on the work that the horse does.

The average cost a year of keeping a horse in Minnesota for the years 1904 to 1907, varied from $75.07 at Halstad, to $90.40 at Northfield.* Often a large number of horses are kept because the farm has been devoted to grain raising and the horses are needed at seeding and harvest times. They run in the pasture during the summer and are idle during the winter months. They must be fed and cared for during this time. The money invested in them would be drawing interest if invested somewhere else. Farmers should reduce the horses kept to the number actually required to do the work, unless the surplus are colts growing in value as one of the market products of the farm. The work of a farm can often be lessened by adopting a good crop rotation and using such crops as do not demand large amounts of horse labour at the same time. In this way a farm of 160 to 240 acres can often be worked with four to six horses, whereas eight to ten are frequently kept. The support of two or three extra horses per year would amount to $200 to $250 and is an item well worth saving.

Investment in Buildings

Investment in buildings, fences, and other items of equipment should be considered in the same business-like way. A barn costing $4,000 and providing shelter for forty head of cattle would carry with it an annual cost of $440. This annual cost is made up from the interest on the investment, insurance, depreciation, paint, and repairs. It will be about 11 per cent on the total investment. If the same forty cattle could be housed in a barn costing $2000 the total annual cost would be only $220, charging the same interest and expense rates as in the first instance, and assuming that the rate of depreciation would be the same. While the $4000 barn would undoubtedly be a better barn, it would not add to the production of the cows housed, unless it was much more comfortable. It would not add to the net profit from the investment unless the labour of doing the chores and caring for the cows would be considerably reduced by greater convenience. The cost of horse barns, swine barns, and other buildings can be similarly calculated. One should not erect a building unless it is going to add to the efficiency of the live stock, shelter hay or machinery, or lessen the labour of doing the chores. It is wiser to invest money in drainage or in better tillage of the soil, than to invest it in buildings that shelter unproductive stock, or that add nothing to the earning power of the farm.

Cost of Labour

The employment, organization and direction of labour demands considerable study. The value of a good farm manager lies quite as much in his abilitiy so to select and direct labour as to yield a profit, as it does in his ability to drive a good bargain or sell his crops well. The only reason for employing labour is to increase the product and consequent profit. If a farmer can, by employing a man eight months in the year at $40 per month, increase the product of his farm by $500, he will be warranted in employing the labour. If, however, the $320 invested in labour should yield an increase of only $200 in the products of the farm, employment would be at a loss.

The Factors of Production

Three primary factors are necessary in agricultural production. These are capital, land, and labour. The adjustment of these three factors is an important part of the business of the farm owner or manager, and determines largely the profits that may be made from the individual farm.

Capital, as commonly understood, includes the money value represented in the investment of the farm property, no matter what the form may be. Implements, live stock, teams, buildings, and other articles of equipment, are each a part of the capital of the farm. Cash for operating is also included.

Land represents the larger part of capital on most farms, and demands special consideration because the amount of land available for agricultural purposes is limited. Farmers have for this reason, regarded it wise to secure large quantities in localities where it was cheap, anticipating a rise in value. Location and demand for land in particular sections has led to much speculation, and land values fluctuate frequently.

The proportionate investment in each of the three forms,—circulating capital, land, and labour—bears a vital relation to the profits possible from the farm, and must be given the most careful consideration by the person who is buying and equipping a farm.

Capital Classified

There are two forms of capital in common use. They are known as fixed or invested capital, and circulating or working capital. The fixed capital properly includes all forms of permanent equipment, such as investment in land, buildings, implements, teams, and other articles that are used continuously. In land it includes the natural value and the value of the improvements that have been made upon it.

Picking stones from a rough section of land adds to its value and increases the capital invested. Clearing trees from the land has the same effect. Wells, drainage, roads, fences, and other forms of improvement which are permanent and which become part of the land, also add to the natural value and become a part of the fixed capital.

Buildings also are looked upon as part of the fixed capital. Strictly speaking, only those buildings which add to the producing power of the farm should be included in the capital invested in the farm. The dwelling house, while commonly added to the investment in the farm, is really intended for the personal use of the farmer and his family. Except in so far as it shelters the help employed on the farm, it can add but little to the returns from it. So far as making a statement of the business of the farm is concerned, it would be better were the farm-house inventoried separately from the other buildings and regarded as a personal expense to the farmer, just as the house of the banker in the city is separated from the business of the bank. All other buildings including silos, corn cribs, granaries, and buildings for sheltering the stock and necessary in conducting the farm business, should be included in the inventoried capital of the farm.

Equipment in the way of teams for work purposes; implements; live stock, such as cows, brood sows, sheep and poultry, that are kept for live stock products, are all a part of the permanent equipment, since they are permanently employed and if sold are replaced by other animals for the same purpose. The circulating or working capital, includes such items of equipment as are frequently changing.

Seed grain, household and farm supplies that are immediately used or marketed, live stock, such as fattening steers, and money for hired labour, are examples of circulating capital. The classificiation intends that the term "circulating capital" shall include only those items that are used once and disappear. If sold for cash, the cash may be invested in other forms of working capital which in turn disappear. Needless to say, the amount of working or circulating capital varies greatly in accordance with the type of business done, with the market, and with the tastes of the farmer. No rule can be given for the exact adjustment of capital for these reasons.

ACCENT ON EMPOWERING THE SMALL FARMERS

Contributions of small holders in securing food for growing population have increased considerably even though they are most insecure and vulnerable group in the society. The off-farm and non-farm employment opportunities can play an important role. Against expectation under the liberalized scenario, the non-agricultural employment in rural areas has not improved. Greater emphasis needs to be placed on non-farm employment and appropriate budgetary allocations and rural credit through banking systems should be in place to promote appropriate rural enterprises. Specific human resource and skill development programmes to train them will make them better decision-makers and highly productive. Human resource development for increasing productivity of these small holders should get high priority. Thus, knowledge and skill development of rural people both in agriculture and non-agriculture sectors is essential for achieving economic and social goals. A careful balance will therefore need to be maintained between the agricultural and non-agricultural employment and farm and non-farm economy, as the two sectors are closely inter-connected.

Raising agricultural productivity requires continuing investments in human resource development, agricultural research and development, improved information and extension, market, roads and related infrastructure development and efficient small-scale, farmer-controlled irrigation technologies, and custom hiring services. Such investments would give small farmers the options and flexibility to adjust and respond to market conditions.

For poor farm-households whose major endowment is its labour force, economic growth with equity will give increased entitlement by offering favourable markets for its products and more employment opportunities. Economic growth if not managed suitably, can lead to growing inequalities. Agrarian reforms to alleviate unequal access to land, compounded by unequal access to water, credit, knowledge and markets, have not only rectified income distribution but also resulted in sharp increases in productivity and hence need to be adopted widely. Further, targeted measures that not only address the immediate food and health care requirements of disadvantaged groups, but also provide them with developmental means, like access to inputs, infrastructure, services and most important, education should be taken.

Identification of need-based productive programs is very critical, which can be explored through characterisation of production environment. We have to develop demand-driven and location-specific programs to meet the requirements of different regions to meet the nutritional security of most vulnerable population in the rural areas. Improved agricultural technology, irrigation, livestock sector and literacy will be most important instruments for improving the nutritional security of the farm-households. Watershed development and water saving techniques will have far reaching implications in increasing agricultural production and raising calorie intake in the rainfed areas. Livestock sector should receive high priority with multiple objectives of diversifying agriculture, raising income and meeting the nutritional security of the poor farm households.

Need based and location-specific community programs, which promise to raise nutritional security, should be identified and effectively implemented. Expansion of micro credit programmes for income-generation activities, innovative approaches to promote family planning and providing primary health services to people and livestock and education should enhance labour productivity and adoption of new technologies. Development of the post-harvest sector, co-operatives, roads, education, and research and development should be an investment priority. A congenial policy environment is needed to enable smaller holders to take the advantage of available techniques of production, which can generate more incomes and employment in villages.

For this poor farmer needs the support of necessary services in the form of backward and forward linkages. Small-mechanised tools, which minimise drudgery and do not reduce employment, but only add value to the working hours are needed to enhance labour productivity. Special safety nets should be designed and implemented for them. Can agricultural co-operatives internalise and galvanize these marginal and excluded people? Off-farm employment provided through co-operatives will go a long way in pulling them out of the state where poverty breeds poverty. Therefore, investment in the empowerment of the small landholders will pay off handsomely. Let us create rural centres of production and processing by masses through co-operatives or empowerment of Gram Panchayats to promote co-operatives. This will improve efficiency of input and output marketing and give higher income. There is need to disseminate widely post-harvest handling and agro-processing and value addition technologies not only to reduce the heavy post-harvest losses but also improve quality through proper storage, packaging, handling and transport. Panchayati Raj institutions and co-operatives can play significant role in all these directions. Giving them power over the administration, as contemplated under the 73rd and 74th Amendment of the Constitution has not been implemented seriously so far in any of the states.

Disaster Management

The frequency and intensity of disasters such as floods, droughts, cyclones and earthquakes have increased in the recent years. The devastating earthquake in Gujarat has brought untold miseries to the whole state and caused a national disaster. Special effort should be made to develop appropriate technologies for increasing preparedness to predict and to manage the disasters. Effective and reliable information and communication systems, contingency planning and national and international mobilization of technologies and resources are a must. Experiences of other countries in prevention and management of the disasters should be shared.

Keeping Pace with Globalisation

The globalization of agricultural trade will bring to the fore access to markets; new opportunities for employment and income generation; productivity gains and increased flow of investments into sustainable agriculture and rural development. I believe that if managed well, the liberalization of agricultural markets will be beneficial to developing countries in the long run, It will force the adoption of new technologies, shift production functions upwards and attract new capital into the deprived sector. However, this will only come to pass if we are mindful of the interests of billions of small and subsistence oriented farmers, fisher folk and forest dwellers in the short and medium tern.

So far the magic of globalization has not been felt in India. During the past one-decade of liberalization certain trends such as deceleration of the growth rate of agricultural GDP, declaration in yield growth rates, and low non-agricultural employment have emerged against expectations. As we globalize, however, it is imperative that we do not forget social aspirations for a more just, equitable and sustainable way of life. Trade agreements must be accompanied by operationally effective measures to ease the adjustment process for a small farmer in developing countries.

Exploiting Cyberspace

Information is power and will underpin future progress and prosperity. Efforts must be made to strengthen the informatics in agriculture by developing new databases, linking databases with international databases and adding value to information to facilitate decision making at various levels.

Development of production models for various agro ecological regimes to forecast the, production potential should assume greater importance. Using the remote sensing and GIS technologies, natural and other agricultural resource should be mapped at micro and macro levels and effectively used for land and water use planning as well as agricultural forecasting, market intelligence and e business, contingency planning- and prediction of disease and pest incidences.

FARM MANAGEMENT

Farm management deals with the organisation & operation of a farm with the objective of maximizing profits from the farm business on a continuing basis. The farmer needs to adjust his farm organisation from year to year to keep abreast of changes in methods, price variability & resources available to him.

Thus farm management is the science which deals with the analysis of the farming resources, alternatives, choices & opportunities within the framework of resource restrictions & social & personal constraints of farming business. This complex information is integrated and synthesized to increase profitability of the farming business, the ultimate aim being to raise the standard of living of the farming people. This does not mean that farm management deals exclusively with the maximization of income; in fact, it takes into account the goals and objectives of the individual farmer, other than income maximization. Thus this discipline deals with people or organisers and decision-makers in respect of farms and agricultural production. It is people-oriented rather than crops or livestock per se.

Farm management is a decision-making science. It helps to decide about the basic course of action of the farming business. The basic decisions of the farming business are:

(a) What to produce or what combination of different enterprises to follow?

(b) How much to produce and what is the most profitable level of production?

(c) What should be the size of an individual enterprise, which, in turn, will determine the best overall size of the farm business?

(d) What methods of production (production practices or what type of quality of inputs and their combination) should be used?

(e) What and where to market?

3

Biotechnology and the Control of Farming

If the only effect of biotechnology and the contract system of guaranteeing property rights were to extend the domain of manufactured inputs into farming, nothing very revolutionary would have occurred. Farmers for a long time have been the purchasers of manufactured inputs. The major structural changes that are occurring in agriculture arise from a vertical integration of farm production in such a way that the purchasers of farm outputs take control of the entire production process.

This vertical integration is made possible by 1) a technical linking of the inputs and outputs, 2) the dual function of a single capital enterprise as both the monopsonist (near monopoly) purchaser of outputs and the provider of critical inputs, and 3) a contract mechanism that links farmer into the loop of inputs and outputs. The use of such contracts predates biotechnology.

Wherever the purchaser of farm outputs is also the processor of those outputs for the market, the possibility of vertical integration has existed. Contract farming has been a common feature of vegetable production for canning. Tomato canneries in Ohio were built in a location central to the farms, the canning company provided the seed and chemical inputs and collected the ripe tomatoes. The farmer provided the land and labour. But the system has evolved greatly since the first canning contracts. The critical role played by biotechnology has been in the material linking of inputs and outputs. In order to guarantee an efficient integrated system of production, the biological inputs into the chain of production, the organisms being raised, are engineered to fit the package of other inputs, the mechanics of the farming process, and the qualifies that the final output is to have for the market.

While some of these aims can be accomplished by conventional methods of breeding organisms, many of the needed qualifies, such as specific disease resistance or qualitative changes in the composition of the organism, are best produced by biotechnological manipulations. Moreover various cloning and cell culture techniques make it possible to reduplicate large numbers of input organisms with desired heritable qualifies, no matter how those qualities were originally produced.

An example of the nature of contract farming is in the production of broilers (chickens raised as meat) where the system is especially entrenched. A major supplier of chickens to supermarkets and fast food restaurants is Tyson Farms of South Carolina. Tyson chickens are produced, not by Tyson "Farms," but by small farmers, owning about 100 acres, producing an average of 250,000 chickens per year, with a gross income of about $65,000 and a net of around $12,000.

This production is under a four-year contract with Tyson (or other similar regional firms), a contract that makes Tyson the sole provider of the chicks to be raised, the feed, and veterinary services. The company is also the sole determiner of the number, frequency, and type of chicks provided. Tyson then collects the mature birds after seven weeks, at a date and time of their own determination, providing the scales on which the birds are weighed and the trucks to take them away.

The farmer provides the labour, the buildings in which the chicks are raised, and the land on which the buildings stand. The detailed control of inputs and farming practices are entirely in the hands of Tyson. So, "The Producer (farmer) warrants that he will not use or allowed to be used... any feed, medication, herbicides, pesticides, rodenticides, insecticides or any other items except as supplied or approved in writing by the Company." Moreover, the farmer must adhere to the Company's "Broiler Growing Guide" and a failure to do so puts the farmer into "Intensified Management" status under the direct supervision of the Company's "Broiler Management and Technical Advisor."

The chicken farmer has ceased to be an independent artisan, buying materials, transforming them by his or her labour, and selling the product on a market. The contract farmer buys nothing, sells nothing, nor makes any decisions about the physical process of transformation. The farmer does own some of the means of production, land and buildings, but has no control over the labour process or over the alienated product. The farmer has then become the typical "putting out" worker characteristic of the first stages of capitalist production in the seventeenth and eighteenth centuries. What the farmer has gained is a more stable source of income, at the price of becoming an operative in an assembly line. The change in the farmer's position from an independent producer, selling in a market with many buyers, into a proletarian without options, is reflected in the nature of the recommendation in the 1998 report of the National Commission on Small Farms, that Congress should amend the AFPA [Agricultural Fair Practices Act] to provide the USDA with administrative enforcement and civil penalty authority that will, in turn, enable growers to organize associations and bargain collectively without fear of discrimination or reprisal. (emphasis added)

The combination of biotechnological manipulation and contract farming also can have a catastrophic effect on third world economies. Much of the import of agricultural products from the Third World consists in qualitatively unique materials like coffee, flavorings, essences, and food oils with special properties. Moreover, the production of these materials is at a low technological level with high labour inputs, in countries with unstable political and economic regimes. As a result, the price and availability of, say, palm oil from the Philippines are unstable. These characteristics make such agricultural products prime targets for gene transfer into domestic species which will then be grown as specialty crops under contract to processors. Calgene has engineered a high lauric acid canola (rape seed) strain for oils that are used for soaps, shampoos, cosmetics and food products that formerly required imported palm oils. These special canola strains are now produced in the Midwest under contract, displacing Philippine production on which a large fraction of the rural population depends economically. And the genes for the biosynthesis of caffeine have been successfully transferred to soybeans. If the essential oil genes for coffee flavor can also be transferred, then Central and South America and Africa will lose their market for beans destined for powdered coffee.

It would be a mistake to think that agriculture has followed the classical picture of the spread of capitalism. Unlike in industrial production, the first step in the capture of agriculture by capital was the immense flowering of input industries and output processors, who appropriated the surplus in agriculture by selling the petty entrepreneurial farmer what he needed, and buying what he produced. No parallel exists in the industrial sphere. It is only with the saturation of that possibility of appropriation that wholly new techniques have come into play. By concentrating on the central material link in farm production, the living organism, which at the same time was the most resistant to capitalization, biotechnology has accomplished two steps in the penetration of capital. First it widened the sphere of input commodity production by including a wide array of organisms that had previously escaped. Second, and more profound, it is making vertical integration possible with the accompanying proletarianization of the farmer. It is this second stage that is the capitalist agriculture of the future, because the physical nature of farm production, inevitably tied to the land, is such as to maintain its unique organization as a productive process.

BIOTECHNOLOGIES FOR LIVESTOCK PRODUCTION

The application of biotechnologies for animal production is more recent in Latin America than for plant production. In 1988, following the recommendations of the First Meeting of the Livestock Development Commission for Latin America and the Caribbean, FAO initiated a series of activities to strengthen institutions in the region working with biotechnology applied to animal production and health, making research contracts with some institutions, in order that they could be better equipped and could function in the medium term as trainers in advanced technologies in the region. In the same year, FAO sponsored the Technical Cooperation Network on Animal Production and Health and carried out biotechnology activities in seven countries of the region: Argentina, Brazil, Colombia, Chile, Cuba, Mexico and Uruguay. The various national institutes in these countries have undertaken specific activities using biotechnologies; these mainly involve: manipulation of DNA monoclonal antibodies, in vitro fertilization, embryo manipulation, immunoassay with radio-isotopes or enzyme-linked immunosorbent assay (ELISA).

To understand the animal biotechnology development in the region, it is important to mention that until the 1980s the industrialized countries generated and developed the majority of the advanced technologies. However, in that decade, Latin American and Caribbean countries underwent a serious economic crisis that prevented them from making the necessary investments so that research institutions and scientific services in support of livestock development could maintain their rate of development and purchase the equipment and instruments demanded by these new technologies. The economic crisis also prevented or reduced programmes for updating and training of scientific personnel.

In view of the above, the majority of scientific institutions in support of livestock development in Latin America and the Caribbean, such as research institutes, schools and facilities of veterinary medicine, schools and faculties of zootechnics, and veterinary diagnostic laboratories, are the only users of modern biotechnological products, which are produced in industrialized countries. Very few of them undertake basic research that will generate knowledge and new technologies in this field. Some institutions with greater biotechnological development do however exist in some countries in Latin America. The most important developments in these countries are indicated below: The Centre of Research in Veterinary Sciences of the Institute of Nutrition and Food Technology (INTA), Argentina, has conducted research into disease diagnosis involving technologies for the production of immunogens through genetic engineering; the expression of FMD viral antigens in eukaryote vectors; the production of Brucellosis antigens through genetic engineering; the development of diagnostic methods for cattle diseases through nucleic acid hybridization; and the production of monoclonal antibodies for rotavirus, bovine herpes virus and FMD virus. The institute provides training in Babesiasis immunology, recombinant vaccines and the vectors used for their preparation. Letters of understanding have been signed with organizations to establish support units to improve dairy cattle and strengthen in vitro fertilization research.

The Biotechnology Centre of the State of Rio Grande do Sul in Brazil, and the Federal University of Rio Grande do Sul (UFRGS), are conducting research into vaccines against the Boophilus microplus tick using molecular methods. Molecular studies are also being carried out for the vaccine against Babesia spp through recombinant DNA and characterization of Babesia proteases to be included in the vaccine. Work is also being undertaken in the repeat sequence cloning of DNA bacteria for the diagnosis and identification of parasites. Molecular studies are also being conducted on the disease caused by Echinococcus granulosus and on its antigens.

The VP 1 and VP 3 in the E. cold expression for FMD are being studied, and monoclonal antibodies against local FMD strains are being produced. Monoclonal antibody techniques are being used to characterize viral diseases. CENARGEN recently offered a course in in vitro fertilization for participants from nine countries. Major research is being undertaken in the Institute of Animal Reproduction at the University of Southern Chile, particularly on immunogens and immunodiagnostics, including ELISA tests and monoclonal antibodies for Brucellosis. Financial support has also been provided for predelivery calf embryo production.

Work is being undertaken in Colombia with the Research Laboratory in Veterinary Medicine (LIMV) of the Colombian Agricultural Institute (ICA). The major research work is on the diagnosis of cattle diseases through DNA probes and monoclonal antibodies, immunogens, vaccines with recombinant DNA synthesized through gene suppression and the preparation of antileukaemics and interferons.

The National Institute of Forest, Agriculture and Livestock Research (INIFAP-SARH) in Mexico is working on various vaccines. It is also promoting the establishment of two genetic engineering research centres for pure and applied research into animal production problems. More specifically, work is being undertaken on immunodiagnostics and vaccines through genetic engineering, monoclonal antibodies and ELISA tests.

The Veterinary Research Centre of Miguel C. Rubino in Montevideo, Uruguay, is working on immunodiagnosis and vaccines using ELISA techniques for different diseases. Work is under way with monoclonal antibodies for different agents and the production of vaccines through genetic engineering. Though biotechnology for animal production is recent in Latin America and the Caribbean, considerable progress has been made and there is a clear trend towards future commercialization, which will further boost its development. Of particular note is the limited work being undertaken on the conservation of germplasm, particularly indigenous germplasm. Whereas biotechnologies have been very active in conserving the genetic diversity of plant matter, little has been done to conserve animal genetic diversity. FAO convened a meeting of experts in Rome in early April 1992 to address this issue.

THE TECHNICAL COOPERATION NETWORK ON PLANT BIOTECHNOLOGY (REDBIO)

The International Symposium on Plant Biotechnologies for Developed Countries, held in Luxembourg in June 1989 under the cosponsorship of FAO and the Technical Centre for Agricultural and Rural Cooperation (CTA), recommended that biotechnology be applied in developing countries to improve food crops. It was stated that biotechnology should be part of a multidisciplinary approach for the genetic improvement of useful plants, and that there was a need to identify the priority issues that could be resolved through advanced plant biotechnology in developing countries. As a response, in 1992, FAO created the FAO/AGP Programme on Plant Biotechnology. The main objective of this programme is to facilitate access to plant biotechnology on the part of the national programmes of Member Nations, particularly those of the least developed countries. One of the mandates of the programme is to support regional networks in order to facilitate the transfer of these modern tools.

In order to response to regional needs, REDBIO, sponsored by FAO, was established following a round table held in November 1990 in Santiago, Chile. REDBIO's main objective has been to apply new technologies to crops of social and economic importance in Latin America and the Caribbean. As of April 1994, REDBIO comprised 323 member laboratories in 18 countries of Latin America and the Caribbean. The data bank also contains information in the field of animal biotechnology.

The network aims to provide a permanent forum for scientific discussion, development and advance; to promote the exchange of biological material and information and access to new biotechnologies; to provide training; and to promote complementarily between research in developed countries and the network's laboratories.

The network has a directory of personnel, an inventory of plant biotechnology laboratories, and details of priorities, species under study and technologies enabling it to determine levels of ability and offer a reference service.

The purpose of the classification of biotechnology laboratories is to establish focal points for cooperative projects and programmes. Information mechanisms are also being established, including the preparation and publication of a periodic circular letter; the collection and publication of plant biotechnology protocols; interlinkage of, and access to data banks; and meetings to discuss priority areas in applied plant biotechnology. With regards to communication among REDBIO members, the Brazilian Centre of Bioinformatic Resources has recently established the Brazilian Network of Molecular Biology and Biotechnology (BMBBNET). This electronic mail allows members of REDBIO to exchange information about laboratories, visitors, consultants, scientific events and courses.

In coordination with subregional and national coordinators, REDBIO organized a regional meeting, BIOCILA, held in Caracas, Venezuela, in November 1992, and the First Meeting of Plant Biotechnology in Brazil. The regional meeting was attended by approximately 250 scientists from the region, the United States and Europe. The meeting in Brazil was attended by 407 experts from 95 laboratories as well as participants from Argentina, Belgium, Costa Rica, Cuba, France, Mexico, Paraguay, Peru, Switzerland, the United States and Uruguay.

BIOTECHNOLOGY RESEARCH AND DEVELOPMENT

According to Sasson (1993), the development of plant biotechnologies in Latin America and the Caribbean was justified by the improvement in the capacity of manufacturing products to meet existing needs, maintenance of a competitive capacity on international markets, access to new markets, alleviation of dependence on imported inputs and its effects on the balance of payments, an orientation towards sustainable development and the conservation and use of genetic resources. Countries were advised to incorporate available biotechnologies into their productive sectors, while taking into account investment requirements, international developments and national capacities. Jaffe (1988) analysed 21 companies in the region that were producing and selling biotechnological products. The study revealed that private companies in Argentina, Brazil,

Colombia, Costa Rica, Chile, Mexico, Uruguay and Venezuela are engaged in a commercial application of modern biotechnology for different purposes, including pharmaceutical, agricultural and industrial products. The positive results of this emerging sector are stimulating the participation of large international companies to invest in the region. Some of the leading companies in the region are Agroceres, Bioplanta and Biomatrix, Brazil; Levaplan, Colombia; Palma Tica, Costa Rica; Biotecnica, Chile; Bioquimex and Genesis Integral, Mexico; Semillas Santa Rosa, Uruguay; and Bioagro, Venezuela.

BIOSAFETY ISSUES

At present, almost 73 percent of total biotechnology research has been directed towards the micropropagation of crop species. Projects using genetic engineering account for about 9 percent of research endeavours. After more than 20 years of progressive activity in plant biotechnological research, Latin America and the Caribbean are in a transition stage, where unsophisticated plant tissue culture techniques are expanding in many countries from the experimental to the productive stage. The multiplication of ornamental, horticultural, fruit and industrial crop species, as well as the recently growing adoption and, in some cases, development of some basic technologies such as molecular genetics tools applied to the diagnosis of diseases, genomic characterization and genetic transformation, were being developed in a smaller group of countries, such as Argentina, Brazil, Cuba and Mexico. The release of genetically modified organisms developed in the industrialized countries has been considered to be the most important biosafety problem in Latin America. In early 1992, authorization for field testing of transformed plants was requested in the following countries: Mexico (eight requests), Costa Rica (one), Argentina (three), Chile (one) and Guatemala (one). Only Brazil, Cuba and Mexico had adopted biosafety regulations in the laboratory, all of which were based on those followed by the US National Institutes of Health. Only Argentina and Mexico had mechanisms for evaluating requests for field testing of genetically engineered organisms. Mexico had the widest experience in testing imported and locally developed transgenic plants, while relying on assistance from the U.S. Department of Agriculture. The relevant regulations were being established within existing norms on plant protection and seed marketing. At the end of 1991, Argentina set up an interministerial commission to evaluate requests for testing and releasing genetically engineered organisms. The same evaluation work was effected by an ad hoc commission in Costa Rica and Chile.

SUSTAINABLE OUTPUT

Sustainable modern and biological farming is an elusive and difficult concept to define when one considers the globally rising population and meet its obligations for hunger and health. Generally, biological farming encompasses objectives of maintaining and conserving soil productivity, conserving of conventional and unconventional energy, environmental quality, food quality, resource quality, preserving Bio-diversity and economic viability of suitable agro-systems. Modern farming methods threaten the environment in many ways: soil erosion, water pollution, the destruction of wildlife habitat, deforestation and the deteriorating fertility of the soil itself. These also bring health risks through widespread use of pesticides. As a result, organic methods of cultivation are being revalued, and more widely adopted.

Sustainable production involves a multitude of "need-based" systems ranging from modern tested technology systems for different crops to an organic sustainable production over time and space. Conversions for conventional (inorganic or chemical or modern) sustainable farming systems to sustainable organic farming amounts to low input costs and premiums by virtue of quality and harmless produce. For instance, a relationship exists between farmers and their customers under the cooperation system in Japan who have adopted organic farming. The produce, despite not being so cosmetically attractive and possibly costing more, between three and five million Japanese now regularly buy organic produce for different health reasons.

Within the context of India, although the use of high-yielding varieties has led to an increase in production, it has been at the expanse of increase in chemical use, which, it is argued, is unsustainable. Low production and productivity are the core to problems of Indian agriculture. Despite substantial increases in the use of modern inputs, the yields are much lower than those in advanced countries.

Factors impeding the growth of crop yields in India are inadequate irrigation facilities, poor manuring practices, and obsolete methods of cultivation. The components of the integrated nutrients supply system suitable for easy adopting include crop rotation, green manures and bio-fertilisers. Effective water saving, equity in water sharing and efficiency in water delivery and use are indispensable for sustainable management of available surface and ground water resources vis-a-vis sustainable production. To sustainable growth of agriculture and prosperity of farmers in real concerns, the use of self-grown and produced inputs has to be promoted. These inputs must replace the market bought modern inputs.

Agriculture should not damage the environment, livestock rearing, regional programmes for protection of water quality, reversion to organic farming, reduction of livestock density by increasing the fodder area of farms, action for the protection and management of fauna and flora, training to local operations to adapt ecologically sustainable farming systems in vulnerable areas and regulations on reforestation are must. When environmental protection is the aim, the maintenance of the farming population is essential to mitigate the problem of unemployment in a country like India.

However, environmentally adapted production systems (such as organic farming) may be transformed into a marketing asset if the products can be processed locally and marketed as a special brand. This cannot usually be done by the individual farmer, but must be done by a group of farmers. In Australia organic farms have shown the lowest nitrate content in the drain water.

The published information on the production of US maize and potatoes employing different technologies and energy inputs is an indicative that the organic production of grains, such as maize, requires less fossil energy than conventional crop production.

For some crops, the organic production system may also produce higher crop yields, as demonstrated with the organic maize system. Combining low inputs with environmentally sound technologies clearly requires that farmers have greater knowledge of the management strategies used in the agro-ecosystem than farmers who employ the more heavy-chemical use systems. For this reason those countries that have active plans to reduce pesticides and commercial fertilisers in agriculture are investing heavily in research and supporting extension education.

Keeping in view the importance of biological farming system, the parameters of research and development — the quantitative and qualitative assessment of costs and returns from alternative farming practices, reducing environmental pollution and health hazards; the inter-dependencies between components of the farming system controlled by the farmer, and interaction of these components with physical, biological and social-economic factors; the inter-disciplinary approach to agriculture research based on an integrated farm production units with limited resources; impetus to investigate and practice organic farming, reduced tillage, methods of conservation of soil as well as integrated pest management, with the overall aim of reducing costs and sustaining yields—must be watched carefully in sustaining any farming system and recommending the best of all.

Super Rice in the Offing

A "Super rice" that can give 25 per cent more yield than all previous paddy breeds is all set to enter the market within the next five years. Developed by the International Rice Research Institute (IRRI) in the Philippines, the new variety could break the yield plateau reached by modern high-yielding rice varieties, according to Mr Gurdev S. Khush, who is leading a team at the IRRI to develop the variety. "Field testing is currently going on in China and in the Philippines, "Khush said, adding the super rice was expected to be in the fields by 2003. Khush was addressing a press conference after signing an agreement between the IRRI and the Indian Council of Agricultural Research (ICAR) for collaboration in rice research.

"The IRRI has developed about 400 different varieties of super rice. They will be tested in India in association with the ICAR within the next two years to find out which one suits the Indian soil and climatic conditions best," he said.

The super rice, developed through cross-hybridisation of two aptly chosen parent plants, had a plant architecture that was completely different from all previous plants, Khush said. It had fewer but sturdier stems, all of them bearing panicles with panicle housing 200-250 more grains than previous varieties.

It took 100-130 days for growth and maturation, Mr Khush said. The dark green-coloured plant had erect leaves to increase photosynthetic effectiveness. It had an organised root system that did not allow wastage of water and micro-nutrients needed for plant growth, Khush said. IRRI scientists were currently working to make it disease resistant by hybridising it with other rice varieties that could resist many of the common plant virus and other insects. "We are now improving the super rice by building multiple pest and disease resistance into it, "Khush said.

In normal paddy varieties, cereal to biomass ratio was 30:70. But super rice was capable of giving a cereal to biomass ratio of 50: 50 which meant a significant increase in rice productivity. With ever-increasing population and decrease in land availability and water, rice productivity needs to be increased by 40-50 per cent by the middle of the next century, "he said.

Novel Scheme for Fruit Growers

THE Horticulture Department of Punjab in its endeavour to improve horticulture has come out with a novel scheme. The department proposes to spread its package of practices to the farmers through its horticulture development officers (HDOs). The state has been divided into various "blocks", each headed by an HDO.

Fifteen villages are selected from each block and from these villages 15 farmers are shortlisted who show cooperation and the will to learn and listen. The department adopts 15 trees of whatever fruit the farmer grows (grapevine in the cases of grapes).

The department has also created "estates" in areas where there is concentration of a particular type of fruit or vegetable. For example, Hoshiarpur is a citrus-growing area and by picking up a central village it covers all the citrus orchards within a 10-km radius. The HDO is required to cover all orchards in the vicinity, contrary to the 15 in blocks. The number of trees adopted varies from estate to estate.

The "thrust", which is also the name of the scheme, is on doing all the activities which are recommended by the department on these 15 plants. The logic behind is to show and train the farmer how proper and timely execution of all farming activities makes a difference to his plants and economy. The unique selling proposition of the whole scheme is that all operations are to be conducted in the presence of the HDO concerned. They may include schedules and methodology of spraying, sowing, pruning, thinning of fruit or any other action which is beneficial to the farmer.

The department claims to have successfully exhibited to peach farmers the technique and importance of thinning the fruit. The fruit size and quality improved considerably after excess fruit was shed. It also showed the farmers the benefits of girdling, brushing and dipping of the grape bunches in giberellic acid.

The programme seems to be excellent on paper but much more is required to seek benefits from such actions. Splinter movements like these offer no gain as they fade away with a slightest change in top management.

The department in its cadre has mostly untrained staff who lack modern perspective and new schools of thought. Secondly, the department is devoid of modern tools and implements to run the show.

For example, it does not have in its possession spray equipment, pruning scissors, tractors, etc. which are so important for the implementation of this project. The department will be dependent on the farmer to meet its requirements. Thirdly, it will be very difficult to monitor the undisciplined staff whose commitment level is always doubtful. Moreover, the department has to find ways and methods which shall highlight its performance in the end and certain yardsticks and comparisons have to be drawn through data to prove and convince the farmer that its package and practices are of benefit to him.

The idea is not to criticise the scheme as on the periphery it seems wonderful but there is many a slip between the cup and the lip. The Horticulture Department breaks the shackles of governmental red tape and hire a non-government organisation (NGO) to evaluate the viability of this programme and how best it can be implemented. By engaging a professional NGO the department shall get a correct picture of its ongoing exercise. It shall wilt away the possibility of a biased report and arrest outdated methods of research and development.

Bihar University to Import 20 Litchi Varieties

RAJENDRA Agriculture University has decided to import 26 varieties of litchi saplings to increase production of the popular seasonal fruit in Bihar. University sources said that the Central Government had already allowed the import of the saplings from China and Australia. Sources said the juicy fruit could emerge as a major foreign exchange earner in the Country.

Only two varieties of litchi, Shahi and China, are being produced in the state. The yield of the litchi could be increased 20 times by adopting mixed cropping which was also helpful in maintaining the quality of the soil. Stressing the need for cultivating bee-keeping in litchi orchards, the sources said it could also increase production of honey.

CROP IMPROVEMENT AND PRODUCTION

Virus-free Materials

Vitroplants of potato, sweet potato, Allium species, taro, strawberry, sugar cane, several ornamentals such as chrysanthemum, carnation, Lilium species and gerbera, and fruit plants such as Citrus species, papaya, apple, sweet cherry, pear and grapes freed of or free from viruses are being commercially exploited in several countries in the region, particularly in China, India, Indonesia, Japan, Malaysia, the Philippines, the Republic of Korea and Thailand. Hundreds of thousands of hectares are planted to virus-free vitroplants of potato in the Region with more than doubled productivity, especially in China and the Republic of Korea. For instance, in the Republic of Korea, the widespread use of virus-free planting material produced through in vitro culture techniques increased the national potato yield from 11.9 tonnes/ha in 1980 to 20.3 tonnes/ha in 1986 (Chung, 1989). Subsequently, an in vitro tuberization system was established and became an integral component of the potato seed industry in the country. Production efficiency of the technique is exceptionally high (young, 1993). One of the main problems encountered in this approach is the chance of reinfection during the propagation of virus-free plants. To overcome and manage this problem, simplified diagnostic methods have been developed and are being used to detect viruses and viroids.

IN VITRO CULTURE FOR MICROPROPAGATION

Besides the above, vitroplants are being routinely mass micropropagated for commercial purposes in the case of orchids, other flowers and ornamental plants throughout the region, especially in Southeast Asian countries (Zamora and Barba, 1990), oil palm (Malaysia, India, Indonesia, Papua New Guinea), date palm (Iran and likely to be taken up in India and Pakistan), embryo culture of soft coconuts (Macapuno) (the Philippines and Thailand), cardamom (India; Kumar, 1990), eucalyptus (China and India), Chinese fir (China), rattan (Indonesia, Malaysia, the Philippines), and seed sterile triploid water melons (Japan). Somatic embryogenesis is also being commercially applied. Synthetic seeds of rice and vegetables have been developed by the private sector in Japan. Triploid rubber clones of rubber produced through somatic embryogenesis have outyielded diploid standard clones by about 20 percent in China. The country has also standardized techniques for mass propagation of sugar cane seedlings using embryogenic cell lines and multishoot propagation systems (Fan, 1991).

To facilitate reforestation and promote social forestry and agroforestry, India has established two tissue culture pilot plants, one at the National Chemical Laboratory, Pune, and the other at the Tata Energy Research Institute, New Delhi (a private organization) with production capacities of five to ten million propagules/seedlings of elite/plus trees of several important species, such as Eucalyptus tereticornis, E. camadulensis, Tectona grandis, Dendrocolamus strictus, Populus deltoides, Bambusa vulgaris and B. tulda (Mascarenhas, 1991). Micropropagation of teak, rattan, eucalyptus, bamboo and other tree species has been adopted in several other countries of the region.

Doubled Haploid Breeding

The region has played a pioneering role in this field. Indian scientists were first to produce haploids from another culture in the mid-1960s (Guha and Maheshwari, 1964). Haploid induction through anther culture has been used most extensively by Chinese scientists for breeding of not only rice but also of several other crops. The technique was used for: (i) production of new varieties;

(ii) production of inbreds for heterosis breeding; (iii) selection of stable resistant lines against biotic and abiotic stresses; and (iv) purification of male sterile lines. One of the most important advantages of the use of haploids is to reduce the time required to develop and release new varieties. For instance, while the conventional pedigree method of breeding and releasing a pure line rice variety requires about 12 years, the haploid method requires only about five years.

Chinese scientists have succeeded in inducing haploids in about 40 different plant species. They were the first to produce haploids through in vitro culture in wheat, maize, sugar cane, soybean, rubber, grapes and apple. In China, several new varieties of rice, wheat, tobacco, and hot and sweet pepper possessing high yield, superior quality, sometimes tolerance to abiotic stresses such as cold, sometimes early maturity, and resistance to diseases have been released through the use of haploids. The most popular of these varieties are Xin-Xin, Hua-Hau-Zao, Zhong-Hua 8, 9, 10 and 11 of rice; Jing-Hua 1, 3 and 5 of wheat; Haihua 1, 3, 19 and 29 of hot and sweet pepper; and Dan-Yu 1, 2 and 3 of tobacco. These varieties occupied about 1.5 million hectares in 1990 (Fan, 1991), of which the rice and wheat varieties accounted for 800 000 and 650 000 ha, respectively.

Other countries have also used doubled haploid breeding methods for crop improvement. In India, rice lines R4 and R10 derived through this method outyielded the check by 15 to 49 percent in trials conducted during 1984-87. The Republic of Korea released two rice varieties derived from anther culture haploid technique and generally screens annually about 6 000 anther-derived rice lines (Chung, 1989). The technique has also been effective in heterosis breeding of Chinese cabbage. In Japan, several successful varieties have been developed through this technique. Among these are rice varieties Joiku N. 394, Hirohikari, Hirohonami, AC No. 1 and Kibinohana, which are tolerant to cold temperature and are good in taste. A broccoli variety, "Three Main", possessing cold resistance and uniform shape, and a cabbage variety, "Orange Queen", with an orange colour, which are quite popular, were developed by the anther culture method (Ito, 1989).

APPLICATIONS OF MODERN BIOTECHNOLOGIES

A number of studies have drawn attention to the region's biotechnological advantages: its traditional involvement in agriculture and forestry; its vast diversity of species and microorganisms; its abundance of natural resources; a good scientific capability in some countries for medicine and biology in general; and active support for R&D. On the other hand, there are also restrictive factors: limited or inexistent linkages between the academic and productive sectors;

lack of venture capital; lack of political support; the brain drain; the reluctance to engage in long-term commitments; and a shortage of funds in universities and research centres. Moreover, limited funds for the development of biotechnologies have resulted in duplication of effort and have reduced incentives for reaching the critical scientific masses, in terms of salary, keeping up-to-date, publications, scientific meetings, reagents and procurement and maintenance of essential equipment. The potential of modern biotechnology is also being held back by insufficient research into strategic disciplines such as genetics and molecular biology, and by inadequate legal and regulatory measures. Biotechnologies in the region are making notable practical contributions to agriculture, and to a less extent to forestry and the livestock sector.

For the reasons mentioned below, biotechnology has focused more on agriculture, as in the developed countries. Plant biotechnologies have been applied mainly to: (i) the natural fixation of nitrogen by bacteria; (ii) the micropropagation of elite material; (iii) the conservation, exchange and use of plant genetic resources; and (iv) genetic improvement through anther culture, somaclonal variation, protoplast fusion and plant regeneration, and genetic engineering. Livestock activity has focused mainly on: (i) diagnosing disease and preparing kits; (ii) in. vitro fertilization and embryo transfer; and (iii) the production of pathogens for the manufacture of vaccines.

Biotechnology has not been applied to forestry to any great extent for technical reasons and because the species involved are perennials with long life cycles. This situation is universal and biotechnological work on woody species is mainly oriented towards research and the development of protocols for the multiplication of conifers and broad-leaved species; the cloning of rapid-growth plus trees (e.g. Eucalyptus hybrids in Brazil); the search for biochemical markers and the conservation of forest resources. Although recently there have been significant contributions to genetic engineering in forest species in Latin America, this development has still not been applied.

ANALYSIS OF BIOTECHNOLOGIES

A number of surveys and studies have been carried out to determine the role, resources and potential of biotechnology in Latin America and the Caribbean, and to provide a framework to examine potential interinstitutional, regional and international collaboration (Sasson, 1993). Three major studies have been undertaken at different times: the 1986 CIAT Survey, Current Status and Future Outlook of Agricultural Biotechnology in Latin America and the Caribbean, that of IICA in 1988, Analysis of Agrobiotechnologies in Central

America; and FAO's Regional Survey of Plant Biotechnology Laboratories (1989-1990). The latter study examined biotechnologies in use; organisms under study; human, physical and financial resources; and major projects in the area. It also reported on institutional plans for the subsequent five years. These were subsequently considered in the FAO study to determine plant biotechnology progress during the five-year period, though no direct comparison could be made. The 1986 survey was based on 95 questionnaires from five geographical areas in the region: the Andes, Brazil, Mexico, the Southern Cone and Central America and the Caribbean. The survey revealed that some institutions had good potential for developing biotechnology for agriculture. It also brought to light an urgent need for interinstitutional collaboration in research and training, and indicated that agricultural research, development and production needed to be integrated in projects attuned to the needs of each crop or livestock production system.

The most frequent technologies used in the region were those with the greatest short-and medium-term potential (e.g. tissue culture). The plants most frequently studied were those of greatest agricultural relevance; livestock (mainly ruminant) research was restricted to embryo transfer and, to a lesser extent, the bioproduction of vaccines; micro-organism studies focused on molecular research; and some limited work was carried out on forest species. A large proportion of those engaged in biotechnology had a high academic level (about 50 percent) and a high proportion was engaged in cell research. Another important fact was the need to absorb approximately 500 biotechnologists (40 percent postgraduates) within the subsequent five-year period.

The FAO Regional Office for Latin America and the Caribbean (FAO/RLAC) organized a second survey to assess plant biotechnology development. The general objective was to obtain up-to-date information on plant biotechnology development, application and prospects in different institutions of the region. A total of 150 laboratories in 15 countries was selected (Argentina, Bolivia, Brazil, Colombia, Costa Rica, Cuba, Chile, Dominican Republic, Ecuador, El Salvador, Guatemala, Mexico, Peru, Uruguay and Venezuela).

The survey, which was carried out in person, comprised eight sections, including laboratory status, appropriateness for cooperative programmes and ability to handle advanced biotechnology. The survey began in October 1989 and was completed in January 1990. This survey revealed a growing involvement of the private sector (27 percent) compared to 1986 (11 percent). The public sector accounted for 73 percent of this activity, with 32 percent in research and 40 percent in teaching and research.

Table : Distribution and level of staff resources in the 153 laboratories surveyed

Academic level	Research orientation			
	RTe[1/]	RTT[2/]	C[3/]	Total
Ph. D.	118	69	20	207
M.Sc.	89	70	26	185
B. Sc.	193	156	73	422
Non-graduates	102	74	76	252
Total	502	369	195	1,066

[1] Research and teaching

[2] Research and technology transfer

[3] Commercial

The heads of laboratory (16.4 percent of the total) had an average experience of 10.8 years and were specialized primarily in tissue culture and plant physiology, followed by genetics and molecular biology (11.8 and 10.6 percent, respectively). This indicates a greater orientation towards basic biotechnology compared to 1986 (Roca, Amezsquita and Villalobos, 1986).

Table : Areas of specialization of heads of laboratory

Specialization	%
Genetics	11.8
Plant physiology	14.2
Biology	3.5
Biochemistry	7.1
Molecular biology	10.6
Microbiology	5.9
Tissue culture	30.2
Plant pathology	6.5
Improvement	3.0
Horticulture	2.4
Fruit cultivation	2.4
Forestry	2.4

In total, the laboratory staff carried out 1 367 projects, with 723 mainly involving basic technologies applied to food crops.

As in other studies, publication was taken as an indicator of productivity, since it represents the end-product of scientific research. There were 552 publications in the last three years, representing 0.67 percent per professional for the period under consideration. This very low productivity applies to all disciplines throughout the region. Latin America accounted for only 1.14 percent of total world publications in 1984 (BID, 1988). One possible explanation for biotechnology is that 28 percent of the laboratories have been in operation for less than three years and 27 percent make their findings directly available. However, the Region clearly has a limited tradition of producing scientific articles. This is the result of a number of factors: there are only a few more incentives to publish than not to, and advanced scientific journals in Spanish are few and poorly distributed.

Both surveys show that plant biotechnology laboratories mainly fulfil teaching and/or training functions because of their linkages to university institutions or research institutes. This is also supported by the fact that there were 537 theses in the last two years on areas of biotechnology research. It was noted that plant biotechnologies are applied to a wide range of crops, including cereals (8 percent), pulses (4.8 percent), vegetables (8.9 percent), roots and tubers (7.2 percent), fruit (29 percent), industrial crops (10.2 percent), fodder crops (2.9 percent), woody species and ornamental plants (7.7 percent).

A total of 63 percent of the projects carried out in the institutions surveyed involved cell and tissue culture, particularly in micropropagation. There was also a marked interest in genetic engineering and molecular biology which accounted for 8.6 percent of the projects. The conservation of germplasm featured in 10.1 percent, disease diagnosis in 10.1 percent, tolerance to adverse factors in 2.7 percent and protein research in 3.7 percent. Most practical biotechnologies such as plant micro-propagation are, therefore, extensively applied in much of Latin America and part of the Caribbean.

Biotechnologies were used for cereals in 108 projects, including molecular probes, genetic mapping and genetic transformation mainly for maize and rice. Similarly, anther culture and the obtention of haploid plants were mainly carried out for rice. The recognized leader in this particular field is the International Centre for Tropical Agriculture (CIAT) in Colombia. Root and tuber biotechnology was the focus of 347 projects, mostly involving cassava and sweet potato, with an emphasis on in vitro conservation of germplasm, diagnosis of viral diseases and micropropagation. The International Potato Centre (CIP) in Peru and CIAT have played an important role in training and in the transfer of technology and germplasm to national programmes.

Fruit, especially indigenous fruit, clearly has potential for the biotechnological obtention and multiplication of genotypes suitable for production for export and/or manufacture of juices and concentrates. There were 357 such projects, mainly using apex cultures.

Similarly, greater attention is being paid to somatic embryogenesis and organogenesis in tropical fruits, and to disease diagnosis. The use of tissue culture for micropropagation of bananas and plantains is common practice in most Latin American tropical countries, and extremely efficient ventures are found in the region (such as Agribiotecnologia in Costa Rica). There were 59 projects on pulses (beans 71 percent) with a predominant focus on tissue culture (43 percent) and genetic engineering (19 percent).

The FAO survey revealed that the major restrictions to biotechnology were training, budget and the lack of an information and/or research network. Table grades the seriousness of these limitations in the opinion of the heads of the laboratories.

Table : Factors limiting laboratory activities

Limiting factor	% of laboratories
Staff training	61.3
Limited budget	55.5
Lack of an information network	50.8
Lack of equipment	50.6
Lack of scientific publications	49.4
Insufficient staff	45.6
Maintenance of equipment	33.3
Lack of projects	21.5
Inadequate infrastructure	13. 7
Obsolete equipment	4.1

As expected, the laboratories varied considerably in terms of performance and facilities.

Some were outstanding, while 42 percent showed shortcomings in basic training (53 percent) and infrastructure (18 percent). The laboratories potential to develop genetic engineering and molecular biology was assessed on the basis of lines of research, productivity and scientific environment.

A total of 15 percent of the laboratories was classified as highly efficient-excellent institutions developing and applying advanced biotechnologies in priority areas for the genetic improvement of crops.

On the other hand, 25 percent lacked the funds and trained personnel to carry out frontline research. The survey revealed that 35 percent of the laboratories were suitable for cooperative programmes. In contrast, 23 percent were considered unsuitable, while the remainder (42 percent) showed limitations in scientific level, training capacity and/or infrastructure.

The prospects of the laboratories were judged on the basis of their ability to develop and apply biotechnologies for genetic improvement and plant production. Approximately one-third (34 percent) of the laboratories will continue to focus on cell and tissue culture, while 22 percent will pursue genetic engineering activities, mainly gene transformation, gene mapping and characterization using RFLP and PCR.

Laboratory funding was found to be a limiting factor in the development of biotechnologies in the region. In many cases the main funding comes from the institution itself, the government budget and/or international sources .

Self-financing (the sale of services and plants) was the main source of income for 30 laboratories (19.6 percent). It was interesting to note that although only ten laboratories procured most of their funds from the sale of plants, 105 were commercially active with a total production of 26 million plants per year for the region as a whole.

The number of propagated plants is significantly lower than the current and potential demand, so there should be a future increase in plant production and entrepreneurial activity. In this connection, most of the countries in the region offer a number of comparative advantages, such as low labour costs and favourable climate.

Table : Main source of funding for the survey laboratories.

Source	No. of laboratories	%
Institutional	44	28.8
International	42	27.2
Government	30	19.6
Sale of services and plants	30	19.6
Donations	1	8

This survey confirmed the existence of a dynamic body of institutions applying plant biotechnology for the main crops of the region. The laboratories carry out considerable research on production problems (723 projects). Genetic improvement and plant propagation, including that of indigenous species, are being conducted through molecular biology, genetic engineering and cell and tissue culture.

The laboratory staffing (1 066 technicians in 153 laboratories) includes 822 professional staff, 47 percent of which with postgraduate qualifications in areas related to plant biotechnology. There is an urgent need for short-term training and updating in genetic engineering, molecular biology, protoplast culture and molecular techniques for disease diagnosis, as was identified also by Roca, Amezsquita and Villalobos (1986).

The third study conducted by the Instituto Internactional the Ciencias Administrativas (IICA) in 1988 (Jaffe, 1988), refers to 26 Central American organizations involved in agricultural biotechnology research.

Twelve were in the university sector, while the rest were public agricultural research organizations, production companies and international and regional centres. Twelve were in Costa Rica and seven in Guatemala (73 percent of the total).

The distribution of human resources confirms the marked concentration of biotechnology activity in Costa Rica and Guatemala which together accounted for 85 percent of these organizations and 93 percent of researchers with postgraduate qualifications in areas related to biotechnology. As in the rest of the region, most of the researchers were located in the universities and regional centres (79 percent).

The public agricultural research organizations and private companies accounted for only 21 percent of the total and only 5 percent of laboratory staff with postgraduate qualifications. Though the international and regional centres only represent 13 percent of the total number of organizations, they total 31 percent of the staff and 38 percent of the researchers with postgraduate qualifications. There is, therefore, a greater concentration of human resources in the centres than in the universities.

Table details the financial and physical infrastructure resources for biotechnologies in Central America. The universities appear to be relatively well equipped but poorly funded in comparison with the international and regional centres. The public agricultural research organizations are in between the two. It is interesting to note that plant biotechnology research has focused on industrial and export crops. In fact, 18 of the surveyed organizations study these crops, while ten concentrate on staple food crops such as roots, tubers, pulses and cereals.

There are four institutions in Central America with significant biotechnology research, application capacity and the ability to use advanced molecular and cell technologies. On the productive level, there are also a small number of enterprises using simpler biotechnologies that require low capital investment. There is a general shortage of high-level scientists, particularly in the national agricultural research organizations.

Table : Organizations in Central America engaged in biotechnology research

Organizations	Costa Rica	El Salvador	Guatemala	Honduras	Nicaragua	Panama	Tot.
Public agric. research organization	1	1	-	-	3	1	6
University	6	-	5	1	-	1	13
Production company	3	-	-	1	-	-	4
International or regional centre	2	-	2	-	-	-	4
Other	-	-	-	1	-	1	2
Total	12	1	7	3	3	3	29

The Tropical Agricultural Research and Training Centre (CATIE) in Costa Rica has been active in biotechnology training for tropical species through its Biotechnology Unit (Villalobos, 1988). Over 300 researchers received multidisciplinary biotechnology training between 1988 and 1991.

Though over 60 percent came from Central America, this training opportunity was open to all Latin America and the Caribbean. Subsequent follow-up indicates that only some 50 percent of those trained have remained in the field of biotechnology. This is unfortunate for the development of biotechnology, especially in Central America where there is a clear shortage of resources, particularly funds and staff.

The very few companies in Central America using biotechnologies have limited experience, mainly involving plant cloning (e.g. Agribiotecnologia of Costa Rica and Palma Tica), the transfer of bovine embryos and the production of animal vaccines. Biotechnology work in the subregion is generally dispersed, particularly in the universities and the public agricultural research organizations.

There is a clear lack of communication between laboratories in the same institution and between institutions within and outside the country. Failure to rectify this, will result in the repetition of mistakes and unnecessary expenditure. It is also clear that the current systems and policies for research funding are not conducive to interdisciplinary or interinstitutional work.

The international and regional centres in Central America account for much of the R&D, so there is strong subregional dependence upon them for biotechnologies. This is also the case of many university research groups and national institutions, because of their low operating budgets and capital investment. The regional economy is not, therefore, in a position to support the current level of R&D investment and the cost of biotechnology (Jaffe, 1988).

Given the inadequate national science and technology policies, and strategies in the countries of Central America, with perhaps the exception of Costa Rica, it is not surprising that there are no national policies for biotechnology development.

However, there have been two recent initiatives to establish such policies which are essential for the consolidation and development of capacities, and to exploit biotechnologies for general agricultural and economic development.

The surveys undertaken in the region permit a comparative and realistic study of biotechnology trends, priorities and limitations. In this connection, the Technical Cooperation Network on Plant Biotechnology (REDBIO), established following the concerns identified in the FAO survey and previous recommendations, should permanently update information and serve as a mechanism to disseminate and promote interdisciplinary regional work.

Practical applications of biotechnology in the English-speaking Caribbean are relatively new. In collaboration with the ministries of agriculture and research and development institutions, FAO coordinated a yam and cassava development project whose main objective was to reinforce tissue culture laboratories and units for the in vitro propagation of important roots and tubers.

This permitted a significant exchange of information and collaboration among national institutions, the University of the West Indies and the Caribbean Agricultural Research and Development Institute (CARDI). This project and other actions have furthered progress in tissue culture mainly for micropropagation and the production and distribution of disease-free material.

The species studied were selected on the basis of their importance for the region, with a particular emphasis on yam, cassava, sweet potato, banana, plantain, cocoyam and tarot The English-speaking Caribbean has an in vitro germplasm bank in Trinidad and Tobago, as well as micropropagation units in Dominica, Barbados and Jamaica.

These units are supported by regional institutions such as CARDI, and international bodies such as CIAT and CIP. The priorities are mainly the obtention of elite material for subsistence and/or export crops important to the national economies. There have been instances of biotechnology research on protoplast culture and somaclonal variant selection for crops such as jute, cocoyam and taro, but there is a serious shortage of trained staff and a clear need to upgrade equipment and infrastructure.

A conference was held in October 1991 in Kingston, Jamaica, on Tissue Culture Technology for Improved Farm Production, under the sponsorship of the Scientific Research Council. The participants including scientists from the area and guest speakers examined biotechnology potential in the Caribbean. They emphasized the need to participate in existing information networks to join forces in seeking international support for postgraduate training and actively to promote biotechnologies for the benefit of the Caribbean countries.

4

Animal Agriculture in India

Greater number of Indian population live in rural areas and engaged in agriculture. But the growth rate of agriculture is only 1.4 per cent in last two decades whrere as growth rate of West Bengal has remained at 3% for last few years.But over all growth rate of GDP in India about 9%may be considered good in terms of economy.Such low growth rate for gearter number of population engaged in agriculture facing crisis resulting suicide of farmers due to various stress factors related to individual economic crisis specially loan taken but failure to return the debt due to loss of agricultural productivity in many states. Scenario is bit different in the State of West Bengal with increase of purchasing power to the rural population.The success came because of land reform and distribution in legal manner.But the future is bleak if there were no industrialization along with nutritious foods security which has linkage with population health.

The growth rate of agricultural productivity obviously found to be lower with higher GDP but no food insecurity in developed countries.Unlike developed countries growth rates of vegetables,fruits,Meats and Fishes etc. in India are higher comparatively.

So,it is high time for India to develop Animal agriculture going beyond calories only. The State Budget of West Bengal has enhanced to Rs.15 crore 85 lacs for the year 2006-07 from previous year of Rs.11 crore 96 lacs to speed up the Animal resources productivity for income generation by self-employment.10 lacs Chicks were distributed among 14000 groups which has been decided to enhance 20lacs for the current year.There is a separate new department called the Ministry for self-employment,Govt.of West Bengal.The Ministry works like Gramin Bank of Prof.Yunus in Bangladesh.

In 10th Animal Development week Hon'ble Governor of West Bengal opened the Programme at Tamluk in Midnapore district on 15th November2006.The State Animal Resources Development

Ministers,Government of West Bengal and Kerala were also present.A seminar on the role of Panchayet and women participation in Animal Resources Development were organized.On the merit of performances of already developed indicators district prizes were distributed.Best Animal Friends were also awarded.

Table showing the ranking of different districts in Animal Resources Development.

Indicators	Rank-1st	Rank-2	Rank-3rd
Milk production	Nadia	Hooghly	Bardhaman
Egg production	24-Parganas(South)	Murshidabad	Birbhum
Artificial Insemination	Nadia	24-Parganas(North)	Siliguri(Darjeeling)
Animal Treatment	Jalpaiguri	Siliguri	Howrah
Green Fodder cultivation	24-Parganas(North)	Nadia	West Dinajpur
Over all success	Nadia		

The National Seminar held from 1st to 3rd February 2007 held in Kolkata also resolved for overall development of Animal resources in India and a model of Artificial Insemination(A.I) Workers at private level.

15 States of India and 150 delegates participated and inaugurated by Mr.Somnath Chatterjee,Hon'ble Speaker of Parliament,India.

1. Doorstep delivery of A.I with all inputs,services like animal health care,milk marketing facilities,supply of feeds and fodder,training of farmers etc. can only be ensured round the clock through out the year by trained private A.I. workers like Animal Friends.
2. Quality A.I services should be assured to the farmers' doorstep through providing quality A.I inputs and proper training to the Private A.I Workers.
3. Each State should formulate Cattle and Buffalo Breeding policy for development of cattle and buffalo wealth in respective State.Necessary legislative support for implementation of the policy may also be considered.Overall technical monitoring and vigil over maintainance of Breeding Policy in the State should be looked after by the Directorate.
4. Livestock resources mapping should be attempted and done as early as possible and conservation of native germplasm should be taken simultaneously by the state governments.
5. One A.I centre (including private A.I.worker) should be established for 800 breedable cattle /buffalo.There should be one Veterinarian for 6,000 breedable animals i.e 16000 large animals.This pattern is advocated for

every State except for isolated areas in islands,hills,desserts etc.There should have uniformity in all the States in engagement of Veterinarian and Private A.I workers.

6. All the agencies engaged in doorstep delivery of A.I services in the State should procure FS Straws only from one window i.e the State Implementing Agency of the State concerned according to norms fixed under Breeding Policy of the State.Necessary legislative support for implementation of the policy may also be considered.
7. Attempts should be taken to provide to Private A.I.Workers not only out of the A.I done but also emphasis should be given to provide incentives on the basis of calf born followed by identification with ear tag and proper records.
8. There should be uniformity in establishment and maintenance cost of Private A.I Workers in all States.(This has a reference to circular no.Q-13021/3/2006-Agri dated 22-05-2006 of Planning Commission, Government of India).
9. Private A.I Workers may be considered as 'Facilitator'in Self-Help-Group (SHG or SGSY) formation and may be utilized as coordinator in co-operative formation.
10. Private A.I Workers are to be encouraged to form 'Growth Centre'for livestock and poultry development in his area,which in future may also act as Micro-Training- cum-Demonstration unit for the farmers.
11. Heifer rearing scheme and calf rearing scheme requires to be reintroduced by Govt.of India under NPCBB by providing funds to the States to boost the cattle and buffalo development programme.
12. Private A.I.Workers may be encouraged to act as a service providerd covering various activities such as Small Dairy units with fodder plots(to serve as demonstration unit),Agri Business Centre (supply of inputs like feed,fodder seeds),vermin-compost units,milk marketing, insurance,etc.
13. Private A.I.Workers may also be considered for various activities financially supported by Credit Linked Capital and /or interest Subsidy/ Venture Capital Fund Scheme(NABARD).
14. Computerisation network,feed back system,data analysis and evaluation at regular interval should be ensured.The Private A.I.workers are the base level information provider.So special care,technical guidance should be assured to get correct relevant information in time from them.
15. For successful implementation of developmental programme,special

care should be given to combat the reproductive diseases and to reduce the calf mortality.

MAN and Women Population(in million) engaged in different sectors of Rural India

- **Women** **Men**
- Agriculture1,870 75 ,470
- Dairying75,000 5,000
- Fisheries1,000 1,800
- Small-animal husbandry15,000 2,000
- Khadi and village industries1,700 1,990
- Handicrafts540 2,200
- Sericulture800 1,200
- Hand looms2,980 4.480
- Source: Various departments of the Government of India.

India should have taken lesson how developed countries developed animal agriculture. As in USA how animal agriculture developed; steps on the part of India might have a suitable lesson.

Are you a livestock producer? Do you know what you would do if you were notified that there was an outbreak of a foreign animal disease in the United States? An outbreak of a disease such as foot and mouth disease would be devastating to the country's livestock industry, according to James Lane, Ford County Kansas Undersheriff.

"Terrorism is a low risk, high impact issue. We want to focus on the day to day criminal activity that is occurring and realize that this is about risk management," said Lane.

"There are so many factors to take into consideration when setting up a plan like this," Lane explained. "It includes involvement from livestock producers, veterinarians, the health department, trucking companies, packing plants, feedlots, the department of transportation, law enforcement, the media and others."

While many groups may have a plan, Lane stressed that it is important to know how to exercise it and know what to do.

CENTER DEDICATED TO PROTECTING AGRICULTURE

The National Agricultural Biosecurity Center at Kansas State University was established in 2002 in response to Sept. 11. The center is dedicated to protecting the nation's agriculture infrastructure by planning and training producers and industry if a threat would occur.

NABC's initiatives include planning, training, outreach and research activities related to threat and risk analyses, incident response and detection and prevention technologies. The center coordinates agriculture biosecurity activities with federal, state and local agencies as well as other university and strategic partners.

"The exercises that have been done with different agencies are important, but the vital work is done before and after these tests," said Marty Vanier, associate director of the center. "Much planning happens before the exercise. After the exercise the groups involved can see what went right and wrong and what worked and didn't work."

Patrick Webb, National Pork Board veterinarian said that since every production system is different, producers should work with their herd veterinarian on reviewing or developing a current biosecurity and security plan.

Webb said that good biosecurity practices can help reduce the risk of disease introduction and spread within a production system. "It can also help contain a disease that is introduced and help to reduce the risk of further spread outside of a production system," he added.

He said that the National Animal Identification System(ID) would support animal health authorities and the industry in rapidly containing and eradicating an incursion of a foreign animal disease.

"The ID system would also help trace the problem back to the site of origin and would also help keep producers in the loop with timely communications on the situation and receive disease control measures necessary to protect their farms," Webb said.

Jon Caspers, vice president of the United States Meat Exports Federation Board and pork producer from Swaledale, Iowa said that his group knows it is pertinent to be able to track a situation back to its origin and feels that an animal ID program would make that happen much quicker.

"If the country affected by a disease such as FMD, it would be devastating to export markets, therefore the faster we find a problem the easier it is to control it if we know where it started," Caspers added. He also stressed the importance of good border patrols and proper trade and public policy regarding animal health issues.(Jennifer Bremer)

Similarly Animal friend in India will be facilitater and communicator in Animal resources development in India.

GFAR conference 2006 held in Delhi,the private sector delegates observed weakness of private sectors to represent the confererence though there is immense possibilities of growth for private investment in this sector.DFID also observed a strong scope of market in veterinary vaccines with increase in animal resources productivity.Resolved for private participation for Insurance and Financial Institutions.

The Chart showing Foods of Animal Agriculture in India

Milk	92 million tonnes
Meat	5.9 million tonnes
Eggs	44 billion
Poultry Meat	1.8 million tonnes

Veterinarians play an important role in monitoring animal resources'productivity.There are 7,872 Veteinary Hospitals/Polyclinics,17,820 Veterinary dispensaries and 25,433 Veterinary aid centers/mobile dispensaries to look after animal health and assist in disease monitoring and surveillance.

There are about 250 disease diagnostic laboratories for quick and reliable diagnosis of animal diseases.About 26 Veterinary biological centers are for production of vaccine and diagnostics. There are 35 Veterinary Colleges and five Veterinary Universities in India to produce veterinary professionals.These are mostly financed by public sectors.

Private sectors take the services of veterinarians to grow by better participation both in micro(Small and Medium) and macro-businesses development causing overall socio-economic development.Animal welfare issues are important in the production of quality of meat and international Trade.There are laws,bills and orders for food safety and acts on prevention of Cruelty to Animals.

An important sector is ICT4D,Electronic medium likeTVchannels and Print media are growing rapidly by Public Private Participation in livestock sector development in India.

According toD-G Council of Scientific and Industrial Research, Bio-suite of Public sector developed in India was a result of a combined public-private enterprise.Private sector devotes less time and investment for doing original research and it is always outsourced from universities and research institutes.

Meat Trade in India

India has the largest livestocks comprising of 189 million cattle(About 80 million unproductive),96 million buffaloes,60 million sheep,120 million goats,185 million layers and 1,100 million broilers.

Total production of meat 5.9 million tones and 1.8 million tones of poultry meat ranking 5th position in the world.

The contribution of buffalo meat in India 25% which accounts for 1.2% of total world meat production.

Table: Buffalo Meat Production 2003-2006 in 1000 Metric Tonnes

Year	Total Meat Production	Buffalo Meat Production	% of Share
2003-04	5898	1471	24.94
2004-05	5921	1582	
2005-06	5889	1592	

The demand for buffalo meat is high as it is relatively lean with fat content below 20% and has low cholesterol.It is cheapest meat in India because of availability at affordable prices as its demand is low as compared to other meats.

Table: Buffalo Meat Production in some of the Asian Countries(2003)

Country	Total Meat in1000MT	Buffalo Meat in 1000MT	Buffalo Meat as% of Total Meat	% of Animals slaughtered
India	5940	1466	24.7	11
Pakistan	1908	582	30.5	15
China	70898	305	0.4	17
Asia	102176	2846	2.8	13
World	253475	3154	1.2	13

Source of Table-I&II FAO(2004)

The lagest importer of Indian meat is Malayasia followed by the Philippines in the year 2005-06, 82.337 MT of Rs.4,994/-crores and 48,918MT of Rs.2,660/-crores respectively.

So,Private Public Participation in this sector will encourage coverage for food security and better Malnutrition Index by the policy makers utilizing veterinary scientists and huge other menpower exist in India.

AGRICULTURE

WAYS BEEN INDIA'S most important economic sector. In the mid-1990s, it provides approximately one-third of the gross domestic product and employs roughly two-thirds of the population. Since independence in 1947, the share of agriculture in the GDP has declined in comparison to the growth of the industrial and services sectors. However, agriculture still provides the bulk of wage goods required by the nonagricultural sector as well as numerous raw materials for industry. Moreover, the direct share of agricultural and allied sectors in total exports is around 18 percent. When the indirect share of agricultural products in total exports, such as cotton textiles and jute goods, is taken into account, the percentage is much higher.

Dependence on agricultural imports in the early 1960s convinced planners that India's growing population, as well as concerns about national independence, security, and political stability, required self-sufficiency in food production. This perception led to a program of agricultural improvement called the Green Revolution, to a public distribution system, and to price supports for farmers. In the 1980s, despite three years of meager rainfall and a drought in the middle of the decade, India managed to get along with very few food imports because of the growth in food-grain production and the development of a large buffer stock against potential agricultural shortfalls. By the early 1990s, India was self-sufficient in food-grain production. Agricultural production has kept pace with the food needs of the growing population as the result of increased yields in almost all crops, but especially in cereals. Food grains and pulses account for two-thirds of agricultural production in the mid-1990s. The growth in food-grain production is a result of concentrated efforts to increase all the Green Revolution inputs needed for higher yields: better seed, more fertilizer, improved irrigation, and education of farmers. Although increased irrigation has helped to lessen year-to-year fluctuations in farm production resulting from the vagaries of the monsoons, it has not eliminated those fluctuations.

Food-grain production increased from 50.8 million tons in fiscal year 1950 to 176.3 million tons in FY 1990. The compound growth rate from FY 1949 to FY 1987 was 2.7 percent per annum. Overall, wheat was the best performer, with production increasing more than eightfold in forty years. Wheat was followed by rice, which had a production increase of more than 350 percent. Coarse grains had a poorer rate of increase but still doubled in output during those years; production of pulses went up by less than 70 percent. The increase in oilseed production, however, was not enough to fill consumer demands, and India went from being an exporter of oilseeds in the 1950s to a major importer in the 1970s and the early 1980s. The agricultural sector attempted to increase oilseed production in the 1980s and early 1990s. These efforts were successful: oilseed production doubled and the need for imports was reduced. In the early 1990s, India was on the verge of self-sufficiency in oilseed production.After independence in 1947, the cropping pattern became more diversified, and cultivation of commercial crops received a new impetus in line with domestic demands and export requirements. Nontraditional crops, such as summer mung (a variety of lentil, part of the pulse family), soybeans, peanuts, and sunflowers, were gradually gaining importance.

The per capita availability of a number of food items increased significantly in the postindependence period despite a population increase from 361 million in 1951 to 846 million in 1991. Per capita availability of cereals went up from 334 grams per day in 1951 to 470 grams per day in 1990. Availability of edible oils increased significantly, from 3.2 kilograms per year per capita in FY 1960 to 5.4 kilograms in FY 1990. Similarly, the availability of sugar per capita increased from 4.7 to 12.5 kilograms per year during the same period. The one area in which availability decreased was pulses, which went from 60.7 grams per day to 39.4 grams per day. This shortfall presents a serious problem in a country where a large part of the population is vegetarian and pulses are the main source of protein.

There are large disparities among India's states and territories in agricultural performance, only some of which can be attributed to differences in climate or initial endowments of infrastructure such as irrigation. Realizing the importance of agricultural production for economic development, the central government has played an active role in all aspects of agricultural development. Planning is centralized, and plan priorities, policies, and resource allocations are decided at the central level. Food and price policy also are decided by the central government. Thus, although agriculture is constitutionally the responsibility of the states rather than the central government, the latter plays a key role in formulating policy and providing financial resources for agriculture.

Land Use

In FY 1987, field crops were planted on about 45 percent of the total land mass of India. Of this cultivated land, almost 37 million hectares were double-cropped, making the gross sown area equivalent to almost 173 million hectares. About 15 million hectares were permanent pastureland or were planted in various tree crops and groves. Approximately 108 million hectares were either developed for nonagricultural uses, forested, or unsuited for agriculture because of topography. About 29.6 million hectares of the remaining land were classified as cultivable but fallow, and 15.6 million hectares were classified as cultivable wasteland.

These 45 million hectares constitute all the land left for expanding the sown area; for various reasons, however, much of it is unsuited for immediate cropping. Expansion in crop production, therefore, has to come almost entirely from increasing yields on lands already in some kind of agricultural use.

Topography, soils, rainfall, and the availability of water for irrigation have been major determinants of the crop and livestock patterns characteristic of the three major geographic regions of India—the Himalayas, the Indo-Gangetic Plain, and the Peninsula—and their agro-ecological subregions. Government policy as regards irrigation, the introduction of new crops, research and education, and incentives has had some impact on changing the traditional crop and livestock patterns in these subregions. The monsoons, however, play a critical role in determining whether the harvest will be bountiful, average, or poor in any given year. One of the objectives of government policy in the early 1990s was to find methods of reducing this dependence on the monsoons.

Himalayas

The Himalayan region, with some 520,000 square kilometers of land, ranks well behind the other two regions in agricultural importance. Despite generally adequate rainfall, the rugged topography allows less than 10 percent of the land to be used for agriculture. The sandy, loamy soils on the hillsides and the alluvial clays in the region's premier agricultural subregion, the Vale of Kashmir—located in the northwestern part of the state of Jammu and Kashmir—provide fertile land for agricultural use. The main crops are rice, corn, wheat, barley, millet, and potatoes. Most of India's temperate-zone fruits (apples, apricots, cherries, and peaches) and walnuts are grown in the vale. Sericulture and sheepherding also are being undertaken. In the eastern Himalayan subregion, the soils are moderately rich in organic matter and are acidic. Although much of the farming is done on terraced hillsides, there is a significant amount of shifting cultivation, which has resulted in deforestation and soil erosion. Rice, corn, millet, potatoes, and oilseeds were the main crops in the early 1990s. The region also is well known for the tea plantations of the mountainous Darjiling (Darjeeling) area in the northern tip of West Bengal.

Indo-Gangetic Plain

The vast Indo-Gangetic Plain, extending from Punjab to Assam, is the most intensively farmed zone of the country and one of the most intensively farmed in the world. Rainfall, most of which comes with the southwest monsoon, is generally adequate for summer-grown crops, but in some years vast areas are seared by drought. Fortunately, much of the land has access, or potential access, to irrigation waters from wells and rivers, ensuring crops even in years of drought and making possible a winter crop as well as a summer harvest. Wheat is the main crop in the west, rice in the east. Pulses, sorghum, oilseeds, and sugarcane are among other important crops. Mango orchards are common. Other fruits of the subregion include guavas, jackfruit, plums, lemons, oranges, and pomegranates. In the Great Indian Desert, rainfall is scanty and erratic. About 20 percent of the total area is under cultivation, mostly in Haryana and Gujarat states, and comparatively little in Rajasthan. The Indira Gandhi Canal—begun in 1958 as the Rajasthan Canal—was designed to bring water from the north. Progress was slow, and only the first stage was close to completion by the end of the Seventh Five-Year Plan (FY 1985-89). By then, the canal had substantially increased the area under cultivation in Rajasthan, and a new completion date of 1999 is anticipated. The cultivable area is expected to expand further with the development of the canal's second stage during the 1990s. The leading crops of the subregion are millet, sorghum, wheat, and peanuts. Vast expanses of sparse vegetation provide sustenance for sheep and goats. In the late 1980s, dairy farming became important in locations that had sufficient pastureland.

Peninsular India

The east and west coasts, the coastal plains, and the deltaic tracts that extend inland for some 100 to 200 kilometers in Peninsular India benefit from both the June-to-September southwest monsoon and the October-to-November northeast monsoon. Farther inland, as the topography and climate change, so does the pattern of agriculture. The proportion of land under cultivation ranges from about 50 percent along the coastal plain and in the western part of Andhra Pradesh to about 25 percent in eastern Madhya Pradesh. Except in areas of certain developed river valleys, double-cropping is rare. Rice is the predominant crop in high-rainfall areas and sorghum in low-rainfall areas. Other crops of significance along the east coast and in the Central Highlands in the early 1990s were pigeon peas, mustard, peanuts, millet, linseed, castor beans, cotton, and tobacco.

On the Deccan Plateau, deep, alluvial black soils that retain moisture for a long time are the basis for much of the region's output of farm products. However, the region also has many farming areas that are covered by thin, light-textured soils that suffer quickly from drought. Whether a crop is made or lost is, therefore, often dependent on the availability of supplementary water from ponds and streams. About 60 percent of the land in the state of Maharashtra was under cultivation in the early 1990s, less in Madhya Pradesh. About 75 percent of the cropland of the Deccan during this period was planted in food crops, such as millet, sorghum, rice, wheat, and peanuts; most of the remaining cropland was planted in fodder crops.

In the far south of the Peninsula, the area under cultivation varies from about 10 percent in the Western Ghats, to 25 percent in the western coastal tract, to 55 percent on the Karnataka Plateau. Here is the India—the land of spices—that Vasco da Gama and other European navigators came searching for in the fifteenth century. On the Karnataka Plateau, sorghum, millet, pulses, cotton, and oilseeds are the main crops on the 90 percent of the cultivated land that is dry-farmed; rice, sugarcane, and vegetables predominate on the 10 percent that was irrigated in the late 1980s. Coconuts, areca, coffee, pepper, rubber, cashew nuts, tapioca, and cardamom are widely grown on plantations in the Nilgiri Hills and on the western slopes of the Western Ghats.

Land Tenure

Matters concerning the ownership, acquisition, distribution, and taxation of land are, by provision of the constitution, under the jurisdiction of the states. Because of the diverse attitudes and approaches that would result from such freedom if there were no general guidelines, the central government has at times laid down directives dealing with the main problems affecting the ownership and use of land. But it remains for the state governments to implement the central government guidelines. Such implementation has varied widely among the states.

Landholding Categories

India is a land of small farms, of peasants cultivating their ancestral lands mainly by family labor and, despite the spread of tractors in the 1980s, by pairs of bullocks. About 50 percent of all operational holdings in 1980 were less than one hectare in size. About 19 percent fell in the one-to-two hectare range, 16 percent in the two-to-four hectare range, and 11 percent in the four-to-ten hectare range. Only 4 percent of the working farms encompassed ten or more hectares.

Although farms are typically small throughout the country, the average size holding by state ranges from about 0.5 hectare in Kerala and 0.75 hectare in Tamil Nadu to three hectares in Maharashtra and five hectares in Rajasthan. Factors influencing this range include soils, topography, rainfall, rural population density, and thoroughness of land redistribution programs.

Many factors—historical, political, economic, and demographic—have affected the development of the prevailing land-tenure status. The operators of most agricultural holdings possess vested rights in the land they till, whether as full owners or as protected tenants. By the early 1990s, there were tenancy laws in all the states and union territories except Nagaland, Meghalaya, and

Mizoram. The laws provide for states to confer ownership on tenants, who can buy the land they farm in return for fair payment; states also oversee provision of security of tenure and the establishing of fair rents. The implementation of these laws has varied among the states. West Bengal, Karnataka, and Kerala, for example, have achieved more success than other states. The land tenure situation is complicated, and it has varied widely from state to state. There is, however, much less variation in the mid-1990s than in the postindependence period.

Independent India inherited a structure of landholding that was characterized by heavy concentration of cultivable areas in the hands of relatively large absentee landowners, the excessive fragmentation of small landholdings, an already growing class of landless agricultural workers, and the lack of any generalized system of documentary evidence of landownership or tenancy. Land was important as a status symbol; from one generation to the next, there was a tendency for an original family holding to be progressively subdivided, a situation that continued in the early 1990s. This phenomenon resulted in many landholdings that were too small to provide a livelihood for a family. Borrowing money against land was almost inevitable and frequently resulted in the loss of land to a local moneylender or large landowner, further widening the gap between large and small landholders. Moreover, inasmuch as landowners and moneylenders tended to belong to higher castes and petty owners and tenants to lower castes, land tenure had strong social as well as economic impact.

By the early 1970s, after extensive legislation, large absentee landowners had, for all practical purposes, been eliminated; their rights had been acquired by the state in exchange for compensation in cash and government bonds. More than 20 million former zamindar-system tenants had acquired occupancy rights to the land they tilled. Whereas previously the landlord collected rent from his tenants and passed on a portion of it as land revenue to the government, starting in the early 1970s, the state collected the rent directly from cultivators who, in effect, had become renters from the state. Most former tenants acquired the right to purchase the land they tilled, and payments to the state were spread out over ten to twenty years. Large landowners were divested not only of their cultivated land but also of ownership of forests, lakes, and barren lands. They were also stripped of various other economic rights, such as collection of taxes on sales of immovable property within their jurisdiction and collection of money for grazing privileges on uncultivated lands and use of river water. These rights also were taken over by state governments in return for compensation. By 1980 more than 6 million hectares of waste, fallow, and other categories of unused land had been vested in state governments and, in turn, distributed to landless agricultural workers.

Land Reform

A major concern in rural India is the huge number of landless or near-landless families, many of whom are wholly dependent on a few weeks of work at the peak planting and harvesting seasons. The number of landless rural families has grown steadily since independence, both in absolute terms and as a proportion of the population. In 1981 there were 195.1 million rural workers: 55.4 million were agricultural laborers who depended primarily on casual farm work for a livelihood. In the early 1990s, the rural work force had grown to 242 million, of whom 73.7 million were classified as agricultural laborers. Approximately 33 percent of the employed rural workers were classified as casual wage laborers.

Because of the large number of landless farmers and the frequent neglect of land by absentee landlords in the early years of independence, the principle that there should be a ceiling on the size of landholdings, depending on the crop planted and the quality of the land, was embodied in the First Five-Year Plan (FY 1951-55). An agricultural census was conducted to provide guidance in setting such ceilings. During the Second Five-Year Plan (FY 1956-60), most states legislated fixed ceilings, but there was little uniformity among the states; ceilings ranged from six to 132 hectares. Certain specialized branches of agriculture, such as horticulture, cattle breeding, and dairy farming, were usually exempted from ceilings.

All the states instituted programs to force landowners to sell their over-the-ceiling holdings to the government at fixed prices; the states, in turn, were to redistribute the land to the landless. But adamant resistance, high costs, sloppy record keeping, and poor administration in general combined to weaken and delay this aspect of land reform. The delays in legislation allowed large landowners to circumvent the intent of the laws by spurious partitioning, sales, gifts to family members, and other methods of evading ceilings. Many exemptions were granted so that there was little surplus land.

To ensure more uniformity in income, to combat evasion of the intent of the laws, and to secure more land for distribution to the landless, the central government in the 1970s pushed for greatly reduced ceilings. For a family of five, the central government guidelines called for not more than 10.9 hectares of good, irrigated land suitable for double-cropping, not more than 10.9 hectares of land suited for one crop annually, and not more than 21.9 hectares for orchards. Exemptions were continued for land used as cocoa, coffee, tea, and rubber plantations; land held by official banks and other government units; and land held by agricultural schools and research organizations. At the option of the states, land held by religious, educational, and charitable trusts also could be exempted. To protect the states from legal challenges to their land reform laws, the constitution was amended in 1974 to include in its Ninth Schedule the state laws that had been enacted in conformance with national guidelines. Land reform laws enacted after 1974 also were included in the amendment.

By the beginning of the 1990s, all states and union territories, except Goa, Arunachal Pradesh, Nagaland, Manipur, Mizoram, and Tripura, had passed ceiling laws to conform to central government guidelines. In Maharashtra, for example, the revised ceiling law that became effective in 1975 set upper limits at perennially irrigated land, 7.2 hectares; seasonally irrigated land, 10.8 hectares; paddy land in an assured rainfall area, 14.6 hectares; and other dry land, 21.9 hectares. By the early 1980s, about 150,000 hectares had been declared surplus under this act, about 100,000 of which had been distributed to 6,500 landless persons. A 1973 land reform amendment in Bihar set a range of ceilings on holdings for a family of five, from six to eighteen hectares depending on land quality, and offered an allowance for each additional family member, subject to a maximum of one-and-one-half times the holding. Within five years, the Bihar government had acquired 94,000 hectares of surplus land and had distributed 53,000 hectares to 138,000 landless families. Success nationwide was limited. Of the 2.9 million hectares of land declared surplus, nearly 1.9 million hectares had been distributed by the end of the seventh plan, leaving 1 million hectares still to be distributed as of early 1993.

By the early 1990s, nearly all the states had enacted legislation aimed at the consolidation of each tiller's landholdings into one contiguous plot. Implementation was patchy and sporadic, however. By the early 1980s, the work had been completed only in Punjab, Haryana, and western Uttar Pradesh and had begun in Orissa and Bihar. In most of the other states, nothing had been accomplished by the early 1990s. The Sixth Five-Year Plan (FY 1980-84) set a goal for the completion of the consolidation of holdings within ten years, which was not achieved.

In order to protect tenants from exorbitant rents (often up to 50 percent of their produce), the states passed legislation to regulate rents. The maximum rate was fixed at levels not exceeding 20 to 25 percent of the gross produce in all states except Andhra Pradesh, Haryana, and Punjab. The states also adopted various other measures for the protection of tenants, including moratoriums on evictions, minimum periods of tenure, and security of tenure subject to eviction on prescribed grounds only. By the early 1980s, most of the cultivated area had been surveyed and records of rights prepared. In most states, revenue assessment—the tax on land—against farmland had been revised upward in keeping with a rise in farm prices. In several states, steps were taken to associate village assemblies, or *panchayat,* with the maintenance of land records, the collection of land revenue, and the management of lands belonging to government; the results of these efforts have frequently been unsatisfactory.

ANIMAL HEALTH AND PRODUCTION

In animal biotechnology, embryo transfer, use of monoclonal antibodies for diagnostics and production of efficient vaccines, hormonal manipulation for growth and productivity, and production of transgenics have been the main areas of work in the region. Nucleus herd improvement using multiple ovulation and embryo transfer (MOET) in Open Nucleus Breeding Systems (ONBS)

Progress through the conventional approaches of genetic improvement of livestock in the region has been slow when compared with the genetic gains recorded in livestock of North America, Europe and Oceania. To overcome this limitation, MOET in ONBS is being advocated in developing countries as the technique may be especially valuable where the use of field progeny testing and artificial insemination (AI) has largely been a failure because of the need for an established infrastructure, both for field recording and AI.

As summarized by Hodges (1988), a genetically superior nucleus flock or herd established under controlled conditions where testing and genetic selection are carried out. The test group is first established by screening the base population for outstanding females. They are then recorded individually in the nucleus herd and the elite females among them are used by MOET with superior sires to contribute embryos which are carried by the recipient cows from the base population. The resulting offspring are reared and recorded and the males among them are evaluated genetically using the performance of their sibs and paternal half-sibs and their own performance where they exhibit the trait.

From these an elite group of males with high breeding value is selected which is used in the base population for genetic improvement by natural service or AI. ONBS yields higher rates of genetic progress per annum than established progeny testing systems mainly because of the shorter generation interval. It is capable of being adapted to different species in which MOET is available. It also permits control of the mix of genes released to the base population. ONBS is flexible for the introduction of exotic germplasm from other populations, for example semen or embryos from temperate breeds and also for the continuous addition of outstanding individuals screened from the base population. China (Chengdu Fengherangshan Dairy Farm), India (Anand Dairy Cooperatives, Gujrat) and the Republic of Korea (Livestock Station, Suweon) are already practicing MOET in their nucleus herds. Similar work at private dairy farms in the Philippines, Thailand, Pakistan, Indonesia and Malaysia has been established, but it is still slow and needs strengthening (Mukherjee, 1989).

Embryo transfer research and application of the technique in improved production and genetic upgrading of cattle, buffaloes, swine, goats and sheep, in varying degrees, are being pursued in several of the Asian countries. Notable progress has been made in standardization of superovulation, synchronization and embryo collection techniques, evaluation and splitting of embryos, optimization of superovulation responses by laproscopy and ultrasonography, embryo manipulation, splitting and standardization of techniques for in vitro fertilization and sexing. Hundreds of calves using these techniques have been produced in China, India and the Republic of Korea and many have also been produced in Pakistan, the Philippines, Indonesia and Thailand, besides the thousands produced in the three developed countries of the region.

In Japan, ever since the standardization of the embryo transfer technique in 1979, the number of calves produced by embryo culture has steadily increased, from 73 in 1980 to 422 in 1984 and to 3 700 in 1988 (Yamauchi, 1989). Artificial production of identical twins (by slicing embryos in two halves) is also commonly practiced In 1988, 268 pairs of such twins were born throughout Japan. The conception rate has been improving, at around 60 percent for fresh embryos and about 40 percent for frozen embryos in the early 1990s. This technique is most commonly applied to the production of high quality beef cattle, such as Japanese Black.

The in vitro fertilization technique in cattle is also being used. In 1987, the first cattle triplet by in vitro fertilization in the world were born at the National Institute of Animal Industry, a research organization of the MAFF. The Institute has also succeeded in nuclear transplantation in cattle (Yamauchi, 1989).

Since the Asian countries have almost a monopoly on buffalo production, they need special research and development programmes on the species. Embryo transfer technology in buffaloes is much more difficult than in cattle. For instance, on average, five to six transferable embryos per donor are recovered in cattle against only 1.5 in buffaloes. Although some striking results have been obtained, such as the birth of a riverine calf from a swamp recipient in the Philippines, or the birth of the first IVF (in vitro fertilization) buffalo calf in India, the techniques need to be improved considerably to be efficient and commercially viable (Mukherjee, 1989). The inception meeting of the regional FAO/UNDP project, RAS/89/001 "Biotechnology Development Network for Animal Production and Health", held in 1988, recommended that a cooperative project be taken up to study the normal hormonal profile of both riverine and swamp buffaloes. The hormonal profile has to be done under varying field conditions. At the same time, the structure of buffalo ovaries needs to be investigated in detail in various breeds. Since the number of primordial oocytes is very low in buffaloes, it is necessary to introduce an entirely different regime for superovulation.

The inception meeting further recommended that once the problem of superovulation in buffaloes had been sorted out, the situation in regard to other items of interest such as embryo splitting in vitro fertilization, etc. could be undertaken. The network should give high priority to the problems of buffalo embryo transfer and countries should work together to produce the necessary data within a defined timeframe on hormonal profiles, structure of the gonad and possible ovulatory regimes that can be taken up for priming the follicles to be followed up by a superovulatory regime.

5

Farm Management: Development of Modern Agriculture

MEANING OF FARM MANAGEMENT

Farm Management is the application of business principles and the scientific principles of agriculture (as discovered by the chemist, the physicist, the agronomist, the animal husbandmand and other specialists) to the business of farming. The farm manager is the one who plans the farm, arranges the cropping scheme, provides the equipment and actually transacts the business and sees to the operation of the farm. The term is a new one and it is important that students of farming get a clear understanding of its meaning. Because of the close relation of farm management to soils, farm crops, live stock, and other agricultural subjects, people often confuse the terms and gain erroneous ideas of what farm management really is.

Farm management is also related closely to rural economy. There is danger of confusion arising in the consideration of these terms also. Questions which relate to the single farm as a unit or to a group of farms operated under the same organization are matters of farm mangagement. Those which relate to a group of farms operated separately, or to a community, are matters of rural economy. The farmer must meet both classes of questions. To state it differently, those things that have to do with the organization and operation within the individual farm relate to farm management. Those that have to do with the external relations of the individual farm to the group or community, relate to rural economy.

MANAGEMENT OF A FARM

Farm management is different from what is commonly confused with the work of a farm manager who manages a government farm as an agronomist.

His function is normally limited to supervising & handling the day-to-day routine of a farm. It normally pertains to the existing pattern of the resource-use & crops-mix under which only the existing plan is executed, supervised & carried out. An intelligent farm manager may go a little further & look after the farm machinery to keep it going.

Farm management, is however, much more than that. Here we are not just concerned with the distribution of labour & irrigation water for day-to-day operations. The emphasis is on the decision-making function of evaluating & choosing between alternative strategies. A major concern is about adjustments which are more suitable & profitable & about exploring new situations & opportunities for maximization of income & satisfying other goals of a farmer. It is the approach under which the opportunity costs of the various resources are evaluated & adjustments in resource-use & enterprise mix are made to secure higher levels of farm income.

Following this approach, the emphasis on yield (productivity per unit of a resource, mainly land) is not ignored, but the greater focus is on the increasing of farm income through a second business organisation. As a business, farming requires the application of business methods & efficient management. To be an efficient manager, one must keep oneself abreast of developments in new technology, new practises, price trends, & economic outlook. Again a farm management man should identify the constraints in the external environment conditions, which hamper a farmer's opportunities & plans for making adjustments in his farm organisation & render the technically superior production plan economically unattractive to the farmer.

ECONOMY OF LANDLORD FARMING IN THE AREA DESCRIBED

We have cited the evidence of agronomists and farmers to the effect that dairy farming on the landlord estates leads to the rationalisation of agriculture. Let us add here that the analysis of the Zemstvo statistics on this question made by Mr. Raspopin fully confirms this conclusion. We refer the reader to Mr. Raspopin's article for detailed data and give here only his main conclusion. "The interdependence of the condition of stock raising and dairy farming, on the one hand, and the number of dilapidated estates and the intensity of farming, on the other, is beyond question. The uyezds (of Moscow Gubernia) where dairy cattle raising, dairy farming, is most developed show the smallest percentage of dilapidated farms and the highest percentage of estates with highly developed field cultivation. Throughout Moscow Gubernia ploughland is being reduced and turned into meadow and pastureland, while grain rotations are yielding place to multi-field herbage rotations. Fodder grasses and dairy cattle, and not grain, are now predominant . . . not only on the farming estates in Moscow Gubernia but throughout the Moscow industrial district".

The scale of butter production and cheese making is particularly important precisely because it testifies to a complete revolution in agriculture, which becomes entrepreneur farming and breaks with routine. Capitalism subordinates to itself one of the products of agriculture, and all other aspects of farming are fitted in with this principal product. The keeping of dairy cattle calls forth the cultivation of grasses, the change-over from the three-field system to multi-field systems, etc. The waste products of cheese making go to fatten cattle for the market. Not only milk processing, but the whole of agriculture becomes a commercial enterprise. The influence of cheese production and butter making is not confined to the farms on which they are carried on, since milk is often bought up from the surrounding peasants and landlords. By buying up the milk, capital subordinates to itself the small agriculturists too, particularly with the organisation of the so-called "amalgamated dairies," the spread of which was noted in the 70s.

These are establishments organised in big towns, or in their vicinity, which process very large quantities of milk brought in by rail. As soon as the milk arrives the cream is skimmed and sold fresh, while the skimmed milk is sold at a low price to poorer purchasers. To ensure that they get produce of a certain quality, these establishments sometimes conclude contracts with the suppliers, obliging them to adhere to certain rules in feeding their cows. One can easily see how great is the significance of large establishments of this kind: on the one hand they capture the public market (the sale of skimmed milk to the poorer town-dwellers), and on the other hand they enormously expand the market for the rural entrepreneurs. The latter are given a tremendous impetus to expand and improve commercial farming. Large-scale industry brings them into line, as it were, by demanding produce of a definite quality and forcing out of the market (or placing at the mercy of the usurers) the small producer who falls below the "normal" standard. There should also operate in the same direction the grading of milk as to quality (fat content, for example), on which technicians are so busily engaged, inventing all sorts of lacto-densimeters, etc., and of which the experts are so heartily in favour (cf. *Productive Forces*, III, 9 and 38).

In this respect the role of the amalgamated dairies in the development of capitalism is quite analogous to that of elevators in commercial grain farming. By sorting grain as to quality the elevators turn it into a product that is not individual but generic (*res fungibilis*, as the lawyers say), i.e., for the first time they adapt it fully to exchange (cf. M. Sering's article on the grain trade in the

United States of America in the symposium *Landownership and Agriculture,* p. 281 and foll.). Thus, the elevators give a powerful impetus to commodity-grain production and spur on its technical development by also introducing grading for quality. Such a system strikes a double blow at the small producer. Firstly, it sets up as a standard, legalises, the higher-quality grain of the big crop sowers and thereby greatly depreciates the inferior grain of the peasant poor. Secondly, by organising the grading and storing of grain on the lines of large-scale capitalist industry, it reduces the big sowers' expenses on this item and facilitates and simplifies the sale of grain for them, thereby placing the small producer, with his patriarchal and primitive methods of selling from the cart in the market, totally at the mercy of the kulaks and the usurers. Hence, the rapid development of elevator construction in recent years means as big a victory for capital and degradation of the small commodity-producer in the grain business as does the appearance and development of capitalist "amalgamated dairies."

From the foregoing material it is clear that the development of commercial stock farming *creates* a home market, firstly, for means of production—milk-processing equipment, premises, cattle sheds, improved agricultural implements required for the change-over from the routine three-field system to multi-field crop rotations, etc.; and secondly, for labour-power. Stock farming placed on an industrial footing requires a far larger number of workers than the old stock farming "for manure." The dairy farming area—the industrial and north-western gubernias—does really attract masses of agricultural labourers. Very many people go to seek agricultural work in the Moscow, St. Petersburg, Yaroslavl and Vladimir gubernias; fewer, but nevertheless a considerable number, go to the Novgorod, Nizhni Novgorod and other non-black-earth gubernias. According to correspondents of the Department of Agriculture in the Moscow and other gubernias private-landowner farming is actually conducted in the main by labourers from other areas. This paradox—the migration of agricultural workers from the agricultural gubernias (they come mostly from the central black-earth gubernias and partly from the northern) to the industrial gubernias to do agricultural jobs in place of industrial workers who abandon the area en masse—is an extremely characteristic phenomenon.

It proves more convincingly than do any calculations or arguments that the standard of living and the conditions of the working people in the central black-earth gubernias, the least capitalist ones, are incomparably lower and worse than in the industrial gubernias, the most capitalist ones; it proves that in Russia, too, the following has become a universal fact, namely, the phenomenon characteristic of all capitalist countries, that the conditions of the workers in industry are better than those of the workers in agriculture (because in agriculture oppression by capitalism is supplemented by the oppression of pre-capitalist forms of exploitation). That explains the flight from agriculture to industry, whereas not only is there no flow from the industrial gubernias towards agriculture (for example, there is no migration from these gubernias at all), but there is even a tendency to look down upon the "raw" rural workers, who are called "cowherds" (Yaroslavl Gubernia), "cossacks" (Vladimir Gubernia) and "land labourers" (Moscow Gubernia).

It is important also to note that cattle herding requires a larger number of workers in winter than in summer. For that reason, and also because of the development of agricultural processing trades, the demand for labour in the area described not only grows, but is *more evenly distributed over the whole year* and over a period of years. The most reliable material for judging this interesting fact is the data on wages, if taken for a number of years. We give these data, confining ourselves to the groups of Great-Russian and Little-Russian gubernias. We omit the western gubernias, owing to their specific social conditions and artificial congestion of population (the Jewish pale of settlement), and quote the Baltic gubernias only to illustrate the relations that arise where capitalism is most highly developed.

Let us examine this table, in which the three principal columns are printed in italics. The first column shows the proportion of summer to yearly pay. The *lower* this proportion is, and the nearer the summer pay approximates to half the yearly pay, the more evenly is the demand for labour spread over the entire year and *the less the winter unemployment.* The least favourably placed in this respect are the central black-earth gubernias—the area where labour-service prevails and where capitalism is poorly developed. In the industrial gubernias, in the dairy-farming area the demand for labour is higher and winter unemployment is less. Over a period of years, too, the pay is most stable here, as may be seen from the second column, which shows the difference between the lowest and the highest pay in the harvest season. Lastly, the difference between the pay in the sowing season and the pay in the harvest season is also least in the non black-earth belt, i.e., the demand for workers is more evenly distributed over the spring and summer. In all respects mentioned the Baltic gubernias stand even higher than the non-black-earth gubernias, while the steppe gubernias, with their immigrant workers and with harvest fluctuations of the greatest intensity, are marked by the greatest instability of wages. Thus, the data on wages testify that agricultural capitalism in the area described not only creates a demand for wage-labour, but also distributes this demand more evenly over the whole year.

Lastly, reference must be made to one more type of dependence of the small agriculturist in the area described upon the big farmer. This is the replenishment of landlords' herds by the purchase of cattle from peasants. The landlords find it more profitable to buy cattle from peasants driven by need to sell "at a loss" than to breed cattle themselves—just as our buyers-up in so-called handicraft industry often prefer to buy finished articles from the handicraftsmen at a ruinously cheap price rather than manufacture them in their own workshops. This fact, which testifies to the extreme degradation of the small producer, and to his being able to keep going in modern society only by endlessly reducing his requirements, is turned by Mr. V. V. into an argument in favour of small "people's" production! . . . "We are entitled to draw the conclusion that our big farmers . . . do not display a sufficient degree of independence. . . . The peasant, however . . . reveals greater ability to effect real farming improvements" (*Progressive Trends,* 77). This lack of independence is expressed in the fact that "our dairy farmers . . . buy up the peasants' (cows) at a price rarely amounting to half the cost of raising them—usually at not more than a third, and often even a quarter of this cost" (*ibid.*, 71). The merchant's capital of the stock farmers has made the small peasants completely dependent, it has turned them into its cowherds, who breed cattle for a mere song, and has turned their wives into its milkmaids. One would think that the conclusion to be drawn from this is that there is no sense in retarding the transformation of merchant's capital into industrial capital, no sense in supporting small production, which leads to forcing down the producer's standard of living below that of the farm labourer. But Mr. V. V. thinks otherwise. He is delighted with the "zeal" (p. 73, *loc. cit.*) of the peasant in tending his cattle; he is delighted with the "good results from livestock farming" obtained by the peasant woman who "spends all her life with her cow and sheep" (80). What a blessing, to be sure! To "spend all her life with her cow" (the milk of which goes to the improved cream separator), and as a reward for this life, to receive "one-fourth of the cost" of tending this cow! Now really, how after that can one fail to declare in favour of "small people's production"!

THE RELATION OF FARM CROPS TO FARM MANAGEMENT

The farmer must grow crops on his land either to sell or to feed to his live stock. This makes it necessary for him to know how to prepare the land, sow the seed, cultivate the crops and harvest and store the product. In gaining this knowledge he studies the questions from the viewpoint of the agronomist and considers only what will be necessary to provide the best conditions for the growing crop.

The farmer must also decide what crops he will grow. One farmer will want to grow corn, another will grow cotton, and still another will grow hay, wheat or potatoes. Usually he will grow several different crops. He will need to decide how much of each one should be grown. In deciding these questions he will study the questions from the viewpoint of the farm manager.

As an agronomist, the farmer is interested only in how to grow the crop. When the crop is mature and harvested, the problem is completed. As a farm manager he still has to settle the question of what to do with the crop. It may be sold. Would it pay better to feed it to live stock? If so, what kind of live stock will give the best returns for the feed used? These are farm management questions.

Relation of Animal Husbandry to Farm Management

Most farmers grow some live stock. They must know the nature of the animals they wish to grow, the kind of food they eat, how to prepare it for them and when to feed them. They should also know how to select the best animals and to what purpose the various classes of animals are best adapted. In answering these questions they study them from the viewpoint of the animal husbandman and cosider only what is best for the animals.

The farmer must decide, however, what classes of stock he will grow. On some farms dairy cows will be best, on others beef cattle and hogs will be a desireable combination. There are farms on which horses and sheep with some poultry will be preferable. There are also the question of how much live stock the farm will supply feed for, or what to do with the surplus when more is produced than is required for the live stock. In answering these questions they are studied from the view point of farm management.

Relation of Rural Economy to Farm Management

The management of a farm includes the buying and selling of products. It is important that the farmer have a knowledge of markets and of values, and that he know when to buy and sell. Often cooperative marketing offers the best opportunity for disposing of products at a profit. So long as the farmer is concerned only with the business enterprises relating to his own farm, he is within the field of farm management. When the business activities and social welfare of several farms or an entire community are collectively considered, the question becomes one of rural economy. Farm management deals with the problems of the individual farm from the business point of view. Rural economy deals with the problems of a collection of farms or an entire community.

Studying Farm Organization

The student of farming is taught how to grow crops by the agronomist, how to grow live stock by the animal husbandman. He learns how to know insects and combat them from the entomologist, and how to cooperate with his neighbors in selling his crops or his stock, from the rural economist. Finally, from the viewpoint of a farmer, he will need to select a specific farm, make plans for its arrangement into fields, select the kind of stock and crops that it will be best to raise, and determine how much of each should be raised. He must provide the buildings and equipment, estimate the capital required and the labour necessary to operate the farm. The farm is thus organized into business form and its operation is simplified. This art of farm organization is learned from the farm manager.

Meaning Made Clear by Questions. "How shall I grow corn?" is clearly and agronomy question. How to select and grow cattle, is just as clearly a question for the animal husbandman. "Shall I grow corn? or some other crop? or cattle? Having grown corn, shall I sell it? or feed it to cattle? or hogs?" are farm management questions. If questions similar to the above are put to each problem presented there should be no difficulty in distinguishing the viewpoint from which it should be attacked.

Definition of Terms. In the pages following, reference will be made to the terms, "Capital," "Inventory," "Farm Receipts," "Farm Expenditures," "Farm Income," and "Labour Income." An explanation of these terms is given below.

Capital. Economists divide capital into two classes: (1) Fixed capital, including land, buildings and permanent equipment, as teams, implements and live stock for purposes of production; and (2) circulating capital, which includes values in seed, feed, fertilizers, supplies and market stock or crops as well as cash for operating the farm. The term is limited to those forms which disappear in the process of production, reappearing in other materials of similar kind.

There is a tendency on the part of farmers to tie up most of their capital in fixed forms, leaving too little for circulating or operating capital. The Inventory is a detailed list of the farm property with values assigned. An average of the values at the beginning of the year and at the close is considered to be the value of the capital invested in the business. The difference between the opening and closing inventories shows the increase or decrease in the value of the farm property.

Farm Receipts are derived from two sources: (1) direct and (2) indirect. The direct receipts are represented in the cash received from the sales of crops, live stock, live stock products, for labour for others and from miscellaneous sources. The indirect receipts would arise in an increase in the inventory due to permanent improvements or to appreciated values of any of the various forms of property. Values should not be inflated for the purpose of showing a gain in the inventory. Debts cannot be paid on anticipated rises in land values or in values of property unsold.

Farm Expenditures

Items paid for in cash for the support of the farm only can properly be classed as farm expenditures. Items of personal or family expense should not be charged against the farm in determining the profits from farming. A decrease in the inventory is considered as an indirect expenditure. The amount of the decrease should be deducted, with the direct expenditures, from the receipts to show the net farm income. The Farm Income is the difference between receipts and expenses. It is the return for the use of capital and unpaid labour. The farmer and his investment unite in earning the farm income. To learn what is earned by the unpaid labour only, the amount that the value of the investment would earn if placed at interest, must be subtracted from the farm income.

The Labour Income is the amount earned by the farmer alone. To learn what this amount is, the amount earned by the money invested at the local interest rate and the value of the labour of other members of the family at local wages while employed at labour on the farm, must be deducted from the farm income. The balance will be the labour income of the farmer.

TYPES OF FARMING

Types of Farming in the United States

Many types of farming are followed in the United States, each type being adapted to some particular locality, climate or crop. The individual farmer must first settle for himself which type of farming will be most agreeable to him. Some men are attracted by fruit growing, others by vegetable gardening, some by cotton or corn raising, and others by grain raising. Under certain conditions live stock raising also is attractive. Here again there are numerous kinds of live stock, and not many farmers can raise all of them. A choice must be made. It will be necessary to determine the possibilities for marketing dairy products and beef products and a decision will have to be made between cattle raising or horse raising as the main business. Sheep, swine, and poultry raising must also be given consideration and fitted in to the plan of farming when they can be added to advantage. Besides the personal preference, the adaptability of soil and climate and demands of the market, of facilities for marketing goods, and the labour supply, must all be considered. Often the amount of capital that can be invested will determine the type of farming. The equipment for certain kinds of farming is much more expensive than for others. Some types of farming call for large amounts of land and small equipment, while other types demand but little land but extensive equipment. These matters can be studied out carefully and correlated so as to balance the business and permit profitable operation. All of these factors are closely related to the financial returns from any farm.

The soil and climate especially must be studied before determining the type of farming. It would be unwise to try to grow corn on cold and wet ground, or to try to grow spring wheat in the tropical climate of the South. Crops, live stock, and methods must be suited to the locality or they cannot succeed. One who wishes to follow a special type of farming, such as fruit growing, cotton raising, or the production of spring wheat must hunt out the locality where the natural conditions are favourable to such crops. This results in much migration and frequent changes. It is best to study out the conditions necessary and meet them so far as possible before becoming established in a business, as moving is expensive.

Types of Farming Classified

Types of farming are usually classified on the basis of source of income, i.e., whether from wheat, or from corn, or from live stock, or some other form of produce. The type may be classified on several other bases such as: (1) The relation to maintenance of fertility, where it is spoken of as exploitative farming if no attempt is made to maintain soil fertility. The prairie soils of America have as a rule been farmed under the exploitative type of farming until the power of producing has been seriously decreased.

It has been necessary to change the type of farming to build up the soil under some other type less destructive of the decaying plant and animal matter which forms an important part of the plant food. History shows that in most cases, twenty to fifty years of exploitive farming has been sufficient on most soils to reduce the production below a paying basis. (2) On the intensity of land operation, whether extensive, as wheat and flax growing on large acreages on the prairies, or intensive, as adapted to truck growing of various kinds. (3) On the diversity of crops or products, thus we have single crop farming as cotton raising or tobacco growing; and the dominant crop farming, where some one crop is made the leading line of production and is supported by two or more supplementary crops.

Classified as to Source of Income

As before stated, it is usual to classify the type of farming under the source of income. By this it is meant that the line of production which brings in the major portion of the income determines the type of farming. If more than fifty per cent of the income from a farm arises from the sale of dairy stock or produce, it would be known as a dairy farm. If more than fifty per cent comes from wheat raising, it would be a wheat farm; or if from cotton growing, it would be classified as a cotton farm. A discussion of the advantages of some of the types of farming as classified under the source of income, follows.

Vegetable Gardening

There are several types of farming that are properly classified under crop growing. Truck farming looking to the production of vegetables is one of the most common. This form of farming must be intensive, and because it is usually necessary to locate a truck farm in the vicinity of a large city or in a particularly favoured locality, it calls for high capitalization. Because of high capitalization, large amounts of labour are required on a truck farm and land may be limited in amount but must be highly cultivated. This type of farming requires two to ten acres of land per family.

It requires much ability in business management because the marketing problems are numerous and must be handled intelligently. The profits from this type of farming are somewhat uncertain though under favourable conditions they may be large. One of the advantages lies in the quick returns from the capital invested. Many garden crops mature within six weeks to two months from the time they are planted and are usually sold for cash. Gardening does not offer employment in the North throughout the year, and unless a greenhouse or hotbed business can be built up with it, there is often lack of employment for labour through a part of the season. It is not adapted to a new country but follows the development of large markets in the vicinity of large cities.

Fruit Growing

Fruit growing like vegetable growing is an intensive form of farming calling for high capitalization and requiring a large amount of labour per acre. It is adapted to somewhat larger land areas than truck gardening, and a family can handle from five to forty acres and often more where fruit growing is made the speciality. Great care is required in the management of the orchards in most localities. While this type of farming is often urged as particularly remunerative and easy, the opposite is usually the case. No type of farming calls for greater watchfulness, more careful attention to details, or more business ability, than fruit growing. Where the best of care is given it may be made a very profitable business. Like truck farming it is not particularly adapted to a new country, though modern transportation methods have to a large extent improved the facilities for marketing the crop. Apple growing will pay in some places between $50 and $150 per acre.* Under the most favourable conditions it may pay four or five times as much in favourable years.

Crop Growing

The single crop system such as cotton, wheat, corn, or tobacco raising requires much less equipment per acre than either truck farming or fruit growing, and can be carried on in a more extensive way. The average cotton farm in the South contains twenty to thirty acres and will grow, when properly managed, $400 to $500 worth of cotton. It is necessary to use a fertilizer in connection with cotton growing to keep up the production as it is an exploitive type of farming,

Wheat growing in the West, especially in the Plan regions along the Pacific Coast, is another example of single crop farming. There is no limit to the size of the farm that can be handled, as the climate is particularly favourable, the dry season at harvest time permitting the grain to stand for a great length of time without injury. Corn growing in the Middle West and South is another example of extensive single crop or dominant crop growing.

In the vicinity of large cities or markets, hay raising is a common type of farming. Because of the low cost of handling land in hay this is one of the most remunerative types of single crop farming. It also has the advantage of being the least exploitative. Hay land is not likely to wash as badly as cultivated land and the humus content of the soil is less likely to be consumed because the land is not tilled as frequently. Where the hay is sold, manures or fertilizers must be supplied to keep up the production.

Tobacco growing, as followed in Maryland, Virginia, Florida, Georgia, Kentucky and some of the other southern states, is another form of single crop farming that is adapted to small scale farming. Five acres per man is the usual size. Tobacco growing is particularly destructive of soil fertility and great attention must be given to the matter of supplying organic matter. Commercial fertilizers have been used and in fact have been thought indispensable in tobacco growing.

Potash especially is demanded by the crop. Tobacco growing is remunerative when properly conducted, but calls for high grade management and business ability. Rice growing along the Atlantic coast and Gulf states;

sugarcane growing in the Gulf coast states; and hopraising in certain favoured localities, are still other forms of single crop growing that are remunerative under certain favourable conditions. All are more or less exploitive of soil fertility and uncertain, as any single crop system of farming must be.

In most parts of the United States, a dominant crop system is better than any single crop system of farming because if the main crop fails there will be sufficient supplementary crops to provide for a reasonable amount of income or at least provide for a living for the family until another crop can be raised.

Mixed Stock and Crop Farming

In the northern states at least, and in the southern states to a large extent, mixed crop growing and live stock raising will be found advisable. The proper arrangement of crops in rotation aids in the maintenance of production to some extent. If these crops can be fed on the farm to live stock which is adapted to the locality and the market, quite as great profit can be made from the crops by marketing them through the medium of live stock as by selling direct. When the manure from the live stock is returned to the farm, the physical condition of the soil is maintained and the chemical elements of the soil built up to some extent. Such farming allows for the employment of labour throughout the year and provides a steady income.

This type of farming is adapted to large or small farms, the size of the business often being determined by the amount of labour available to care for the live stock. It allows for the selection of a dominant crop or live stock products suited to the demands of the market. While it does not pay so large a margin per acre as some of the intensive single crops, the returns form mixed stock and crop farming are much more certain. Under this system the average farm family can care for the products from eighty to one hundred-sixty acres of land. When well managed, and the corps and live stock properly proportioned, they may handle farms of two hundred forty acres or more in size to advantage. Gross returns from the farms so handled should be $15 to $25 per acre.

Live Stock Farming

In certain sections of the United States live stock has been made the main line of production. Live stock farming includes the production of beef cattle; the production of sheep; hogs; dairy cattle; horses; and poultry. Some conditions call for the production of only one class of live stock. Other conditions will permit the combination of two or more kinds. Stock raising of any kind calls for high capitalization in buildings and live stock equipment. It carries large risks and demands a large amount of labour. This field particularly calls for managing ability on the part of the farmer. Unless live stock is well bought, properly matured and finished, and well sold, there is likelihood of loss. Where good management is given, however, live stock raising is one of the most remunerative types of farming and provides a certain income. Dairying especially is looked upon as providing a sure income when conducted in the vicinity of good markets. With cows of high yielding qualities it is possible with good management, to secure a good income.

PLANNING THE FARM

Need of Planning

A practical, well-balanced farm business cannot be built up without a great deal of careful planning. Among the most important plans to make is one of the whole farm, showing the location of the farmstead and its connection with the fields, also the size and the arrangement of the fields themselves. The ease with which the farm may be operated and the cost of the farm products are largely dependent upon the arrangement of the buildings and fields, and upon the plans for operating the farm. So important is this matter that several different plans should be made before adopting a final one. Often it will be found impossible to make a plan that can be immediately adopted. A tentative plan may be followed for a year or two, while the fields are being subdued and cleared, or it may be necessary to change the type of farming within a few years. These prospective changes, however, should not prevent one from working out the complete plan for future operation.

In fact, the more one works over the plan, the greater the possibility of having all features of the farm in proper balance. Owing to the great difference in the topography of the land used for farming, it is not possible to present a plan which can be adopted by everyone. Planning a farm is like planning a house in that the tastes of the individual family that is to live upon it must be consulted, and the size, shape, and internal arrangements should be determined by the use to be made of it. The principal factors are presented to serve as a guide to those who wish to work out a plan for operating a farm.

Access to Fields

In laying out the farm, the farmstead should be located and planned so as to permit easy access to all of the fields. On most farms the products of the fields are brought to the farmstead for preparation for market or to be fed to live stock. The manures from the barns and barnyards should all be taken back to the fields to aid in maintaining soil fertility. Easy access to all of the fields makes more certain the even distribution of manure over the farm and lessens greatly the labour in hauling in the products and returning the by-products.

The shorter the distance to the fields, the more economically can this be done. A large number of trips are made to and from the fields in preparing the land, and in cultivating and harvesting the crops. If each of these trips is only a few rods longer than is necessary, much loss of time will result. Such loss of time is expensive and increases the cost of producing the crops. As an example of how great this loss may be, attention is called, showing the average distance from the farmstead to the fields on a farm poorly planned and the distances on the same farm, well planned. It is estimated that with the farmstead 45 rods farther on the average from the fields than need be, a distance of 107 miles would be travelled in raising a crop of corn on a 27-acre field. The total trips to and from the field would be 383, which multiplied by 90, the unnecessary distance from the fields, would make 107 miles. This would be equivalent to four days of extra time, which, at the ordinary price for farm labour, would cost $16 to $20.

The cost of growing the other crops would be proportionately as great. One cannot over-estimate the advantage which comes from having fields well planned and closely connected with the farmstead. Provision should be made for the men and teams to begin working immediately upon leaving the buildings.

Shape of Fields

The shape of the fields also has considerable influence on the economy of handling the farm. A long, narrow field is more economically worked so far as the use of the teams and machinery is concerned. Such a field requires less turning which consumes the time of the men and the team. A triangular field is very expensive to work as are fields with irregular outlines. Observations made on the time required to turn with the various farm implements, and calculations as to cost, show that the expense of growing crops on a square field is increased twenty to twenty-five percent by running a ditch diagonally across the land. So far as possible, therefore, with a view to economy in operation, the fields should be made comparatively long and narrow.

As an opposing factor, however, the expense of fencing fields of such shape must be considered. A piece of land one rod wide and 160 rods long, contains an acre. It would require 322 rods of fencing to enclose it. An acre of land square in form would be 12 2/3 rods on each side and would require only 51 rods of fencing to enclose it. The problem of fencing must be balanced against the ease of working the land and a medium between the two reached. Ordinarily the rectangular field, about twice to three times as long as it is wide, works out to the best advantage when both of these factors are considered. The diagrams of an 80 acre farm fenced in each of three different ways, illustrate forcefully the wisdom of careful planning in this regard.

Size and Number of Fields

The size of the fields is also an important factor in planning the farm. So far as possible, the fields should be uniform in size, as uniformity in size leads towards making the farm business more systematic. Such an arrangement of fields provides a definite acreage of each crop each year and enables one to regulate the amount of live stock to be kept. It gives the assurance of a plentiful feed supply each year and a somewhat constant income. The amount of machinery, the work stock for operating the farm, and numerous other factors are also influenced by the size and uniformity of the fields. So far as possible, the fields should be of good size. They must be suited to the rotation that is to be followed on the farm. On many farms it is impossible to change the size or the shape of the field materially.

The topography of the land has much to do with the divisions in the fields. Creeks, rivers, or sloughs, often present unconquerable barriers to the proper division of the fields. Hillsides, rocky ridges, and even sandy bottoms may change materially the desired arrangements so far as the size and shape of fields is concerned. All of these matters must be considered, however, and the ideal followed as nearly as possible. Many people believe that the adoption of a rotation of crops will increase the number of fields on the farm and object to it for that reason. Careful study of hundreds of farm plans presented in the farm management class of the Minnesota Agricultural School show that the adoption of crop rotation and the division of the farms into fields reduces rather than increases the number of fields on most farms.

Rotation Groups

On many farms it is well to have at least two, and in some instances, three groups of fields, suited to different rotations. The major rotation may consist of three to six or seven fields, and should receive the greatest consideration because it covers the larger part of the farm.

It is on the major rotation fields that the main crops of the farm will be grown and where most of the work will be performed. These fields should be comparatively large and the farm so planned that there is a field for each year in the rotation. So far as possible, the fields should be of the same size. On most farms it is possible to have a minor rotation group of small fields. The fields in this group should be located close to the farmstead and intimately connected with it by short lanes, and it is desirable that they be fenced. Minor rotation fields are calculated to grow pasture and early forage crops for the live stock and to supplement the major rotation in the production of certain desirable crops.

On large farms it is sometimes wise to have two or more groups of rotation fields. This arrangement permits of raising double quantities of some of the more staple crops, and of reducing the amount in the minor crops. The arrangement of the farm into groups for rotation purposes.

The Original Plan

The first step in planning a farm is to make an outline drawing, showing the plan of the farm as operated at the present. This original plan should show the natural subdivisions of the land and the topography so far as possible. It should show roads, fence lines, as well as creeks, tile drains, or open ditches, and the location of the farmstead and any groves or trees that may be on the place. Distances should be measured or closely estimated, and as many of the details as possible filled in. This original plan will serve as the basis for future development.

Transition Plans

In the development of the plans of the farm, it is frequently necessary to make one or two transition plans showing the changes from year to year. It is impossible, without loss, to change immediately to the desired plan. Fields will need to be seeded to grass, and those in grass will, it may be, have to be broken. So it takes two or three years to bring about the complete transformation. The transition plan is necessary as showing the record of the crops and the projected plans for the future.

The Permanent Plan

The permanent plan when completed should show the location of the farmstead, the minor rotation fields, and the major rotation fields. It should show also the lanes and roads connecting with these fields; the location of the buildings; arrangement for water supply; and the complete plan of cropping for each rotation cycle. The more complete the plan is made, the less trouble will be experienced in following it. Provision should be made for changing from time to time, as market conditions or the inclination of the owner should demand. Provision should also be made for catch crops and for filling-in when crop failures are experienced. The necessity for filling-in is bound to rise through crop failure, but such failure should not interrupt for any length of time the plans for the permanent operation of the farm.

Crop Ledger Plan

A very convenient form of recording the yield and production of the farm is shown in the crop ledger plan. This consists of an outline drawing, showing the fields and their arrangements. It is planned that one of these maps will be used each year in filling in the crops grown, the yield per acre, and the total yield on each field of the farm. It serves as useful memoranda and when preserved it will often be referred to as a part of the crop history of the farm.

History of Manuring

A record should be made of the date on which each field is manured. Ordinarily, barnyard manure is applied preceding the corn crop, and a record of this could be made in the regular rotation plan. Any deviation from the plan, however, would not be shown, and it seems advisable to use a separate outline map of the farm as a basis for the history of manuring [or fertilizing, ed.] each of the fields. By using the same outlines used in planning the rotations for the farm, the history of manuring each field can easily be designated by "cross-hatching" the particular field manured and inserting the number of the year in which the manure was applied. Where only a part of the field is manured in a single year, that part of the field to which the manure was applied should be "cross-hatched."

Production Limited by the Deficient Factor

It is a common experience that the production on a farm is limited by the minimum amount of the one deficient factor. Difficulty is experienced in securing profitable production on a limited land area. In such a case, labour will not be fully employed or the equipment cannot be used effectively. On the other hand, a large land area and large equipment cannot be used to advantage without a plentiful supply of labour. Again, neither land nor labour can be used to the best advantage if the equipment is inadequate. It therefore stands to reason that these three factors must be carefully considered and proportioned in accordance with the needs of the business.

In purchasing and organizing a farm with limited capital, it is believed best to make the investment in about the following proportions:

- 45% for land investment
- 20% in buildings, provided they are to shelter productive live stock or market products;

- 22% in work animals and live stock;
- 8% in implements and tools;
- 5% reserved as working capital.

In buying a farm and equipping it with new machinery, the land investment will run lower than above indicated and the implements and tools investment will run higher. The machinery, however, will depreciate in value while the land is more likely to increase in value. As the farm becomes older and the land is improved, the proportion of investments will gradually change. On old farms near city markets, the machinery investment may become comparatively insignificant. The proportionate investment will necessarily vary with the type of farming followed. In the highly intensified forms of farming much more will be invested in live stock. H. W. Mumford, in Bailey's Cyclopedia of Agriculture, gives a table showing the investment on a live stock farm in Illinois, from which it is calculated that there is invested in land and buildings, 60 per cent; in machinery, 4 per cent; in live stock, including horses, 30 per cent; and in labour, 5 per cent.

EXERCISES FOR PUPILS

Farm Inventory

Have the pupils take inventories of their fathers' farms, providing a form similar to the following. The inventory may be taken on the regular weekly holiday. Prices should be checked over by the proprietor of the farm and brought to class for discussion and completion. The teacher should, if possible, go with the class to some farm and take an inventory, with the help of the proprietor, before starting the class members on their individual work. Allowing for any withdrawals of or additions to capital, the difference between the totals of two successive inventories will be the gain.

Proportion of Investment

Have the pupils find the total investment and the per cent of the capital invested in each of the forms of equipment used in classifying the inventory.

Cost of Shelter

Have the pupils learn the cost of the barns on their fathers' farms. Ask them to calculate the interest at the prevailing rate. Determine the amount of depreciation on each barn at 5%. Learn the cost of insurance and repairs for the year. Include the interest, depreciation, insurance, and repairs in one sum called the "Annual Cost." Divide this annual cost by the number of animals sheltered and learn the cost of sheltering one head on each of the farms. Where different animals are kept, they may be reduced to a comparative basis by estimating the weight and considering 1000 pounds as a unit. Ten pigs weighing 100 pounds each would equal 1 cow or colt weighing 1000 pounds.

THE FARMSTEAD

Organization

The organization of the farm should begin with the farmstead. It is from the farmstead that the daily operations will be directed. Its location and arrangement are vital to the successful and profitable management of the farm and should be carefully studied. Two opposing factors will demand consideration in making the plans. (1) It is to become the home of the family. Everything that will add to their pleasure, comfort, and happiness should be included. (2) It is the centre of the business operations of the farm, and it is essential that the plans provide for the economical administration of the farm business.

The first factor demands for the farmstead a position near the highway for the sake of social convenience, or near a lake or pleasant landscape view for aesthetic reasons. The second factor demands that the buildings and yards be near the centre of the farm with easy access to the fields so as to permit quick connection with the work and short distances for transporting the farm products to the farm centre. These opposing factors must be harmonized as far as possible and the best location chosen, all things considered. Whatever the location finally decided upon, the economical operation of the farm depends upon the farmstead being well arranged and the farm work being organized into a systematic and definite relationship with it.

Location of Farmstead

In locating the farmstead it is best to choose a place near the centre of the farm when this can be done without sacrificing the social interests of the family. The advantages of having the farmstead located near the centre of the farm are many. Time is economized in getting to and from the fields. All four sides of the farmstead are in connection with the fields, allowing minor rotations to be arranged for, which will provide the live stock with pasture and paddocks close to the building. This permits live stock to be taken quickly from the fields to yards or barns. From a central location the farm manures can more easily be taken to all of the fields of the farm without hauling it long distances. Also, the crops grown on the various fields can be more readily brought to the farmstead. The objection to a central location is that it results in the isolation of the farmer's family, which is a serious phase of farm life. Families of farmers enjoy seeing passing teams and should be near enough to the road to encourage social calls from their neighbors. Convenience of getting to school and to market should also receive due consideration. The economic advantage of being near the centre of the farm, but it is doubtful whether this advantage is sufficient to outweigh the social advantages of living near the main road at the centre of one side of the farm. When the farmstead is properly arranged with short lanes leading to the minor rotation fields, and with the major rotation fields connected by roads, but little of the economic advantage is lost. Both locations should receive careful consideration and the one adopted which gives the greatest comfort to the family and which does not call for useless expense in the operation of the farm.

Size of the Farmstead

The size of the farmstead should be in proportion to the size of the farm. It will often be determined by the nature of the farm business. Live stock raising with suitable barnyards and paddocks demands more land in the farmstead than where grain raising or fruit growing is the principal business of the farm. Four to six acres on a 160-acre farm is not too much for the farmstead on a diversified grain and stock farm. Six to ten acres can often be used to advantage. A plot of ground 32 by 40 rods in dimensions on reasonably level land will give an eight acre farmstead of suitable proportions. The dimensions can be varied in any way desirable, or necessary to suit the particular farmstead to be planned. Room should be reserved for wind-breaks or shelter-belts. The orchard, garden, poultry runs and buildings for live stock, with suitable yards and paddocks, should be included. Provision should be made for drives and walks between the buildings. A generous lawn adds greatly to the beauty and attractiveness of the farm home.

The only objection that can be made to such a lawn is that much labour would be required to keep it up. By a little careful planning a large part of the labour can be avoided.

By grouping the plantings of shrubs, protecting them by light wire hurdles, and by providing permanent fences on certain sides, it is often possible to keep the lawn trimmed by allowing the sheep and calves to graze on the lawn for two or three hours each day. It may be necessary occasionally to run the lawn mower over it to even it up, but ordinarily, where a little attention is given to the grazing of the sheep and calves, they will keep the lawn in good shape without additional labour. The only objection to this plan is that the droppings from the animals are disagreeable. This annoyance can be avoided to some extent by turning the animals on for two hours or so early in the morning when the animals have been off feed over night. By using the orchards for poultry runs or for pig or calf pastures, good use can be made of the land. Other portions of the enclosure may be used for permanent paddocks or stock runs, putting the less objectionable animals in the foreground.

The whole effect of the landscape should be broad and commanding with the shelter-belt as a background, and with suitable decorative plantings and lawns in the foreground. An acre of the land can be used about the house itself to good advantage.

Location of the House

The house should have the choicest position in the farmstead. It should be on high ground, sloping in all directions so as to provide good surface drainage. An east front is considered best and if the house can be located to the south or east of the farm there will be less annoyance from objectionable odors during the summer season when the prevailing winds are from the south and southeast. It should not be located near a slough or pond of stagnant water unless provision can be made for drainage. In starting with a new farmstead, little trouble will be experienced in finding a proper site for the house if the grounds are examined carefully, and if the farmstead is mapped out to scale on paper.

After the paper plan is made, the land should be measured and stakes set for the various buildings, thus establishing the proper relations and distances between them and the house. Provision should always be made for future growth.

Sites should be reserved for buildings which are to be added later, and these buildings should be located upon the paper plan. In this way it will be possible to avoid moving buildings or destroying wind-breaks, orchards, or groves which have been planted in the wrong place. Pleasant views should always be sought from the site of the farmhouse. A lake, a grove, a schoolhouse, or a view in the distance of some city or town, or of a mountain or hill, may be had by arranging suitable openings in the wind-breaks or groves. In like manner, unpleasant objects can be hidden from view by groups of trees. Frequently a building in the farmstead itself may be made to shut out an objectionable view from the house. It is a mistake to plant trees too close to the house, thus shutting out a free circulation of the air. Allowance should be made for the growth that trees make and they should be planted some distance apart, or plans should be made for taking out every other tree as they get larger.

Remodeling an Old Farmstead

In remodeling an old farmstead the problem is somewhat more difficult. Frequently the buildings are improperly placed. It is not always possible to work such a farmstead over into a satisfactory farm home. Groves can be changed, however, in the course of eight or ten years and minor buildings may be moved at small expense, and new buildings erected where they will fit in with the plan as it must be followed. It is always possible to hide objectionable views by clumps of shrubbery or by hedges. Drainage, the filling of low spots, or a little grading about the house and farmstead, will often change an undesirable place into one that is quite satisfactory.

Grouping of Buildings

In arranging the buildings on the farmstead, certain groupings should be made which will greatly reduce the amount of time required in doing the daily chores. The horse barn and machine shed with workshop included, should be placed near together. Both should be located near to the main lane leading to the fields. Consideration of the daily tasks shows that the farmer is required to go to the horse barn for his team; from there he goes to the machine shed for the implement he is to use, and he should then go direct to the field. The horse barn is the starting place for most of the farm operations where power is required. The machine shed furnishing the implements and the workshop should be convenient for the sake of economizing time in keeping the tools and machines in repair. Going from the horse barn across the farmstead for a machine and thence back on the way to the field, causes much unnecessary travelling and should be avoided where possible. On three farms in Minnesota the distance travelled in going to and from the machinery shed and the horse barn varied from 11.3 miles where the buildings were closely connected, to 68 miles per year, where the buildings were widely apart.

The cattle barns and the buildings containing feed supplies, should be located closely together. If the feed can be stored in the same building, so much the better. The silo and hay-mow at least should be closely connected. Grain supplies can be brought from a distance if necessary, but it is better to make provision for them in the barn. A milk room should be provided if possible. It may be combined with a well-house and tank which will supply water at convenient points for the live stock. It is sometimes included in the barn, though there are many objections to this plan as it is difficult to keep the barn odors from reaching the milk. It is best to have the milk room separate from the barn.

It is very essential that the corn crib be closely connected with the hog and cattle feeding yards. A trip of 250 feet across the farmstead and back three times a day for a basket of corn may mean as much as 103 miles of travel per year. This can be avoided by placing the corn crib near the hog feeding floor. Too much attention cannot be given to this matter of suitable connections between buildings, as it bears closely upon the economy of performing the daily tasks about the barns and yards.

The poultry house should, if possible, be given a location near the farmhouse and not far distant from the grain supplies. Frequently the poultry house can be near the orchard. Both are looked after or visited frequently by the women of the family and they, as well as the men, should be saved every unnecessary step.

The orchard can sometimes be used for the poultry run, thus keeping down the insects and making double use of the land. The garden also should be near the house, and it should be fenced to keep the poultry out, unless the poultry is confined in a yard of its own.

Lanes and drives to the fields should be arranged for in the plans of the farmstead. The permanent fence lines should be laid out with this in view, temporary fences being erected when necessary to aid in keeping the farmstead clean and attractive. Main drives and roads to the fields should be open to give easy access. By putting a temporary hurdle or fence across, the corners of the farmstead and the ground adjoining the barns and storage buildings can be trimmed up by turning in the sheep or the colts for grazing a few hours each day. In this way the whole farmstead can be kept attractive and bad weeds subdued.

Time Required for Chores

That the arrangement of the farmstead has a great deal to do with the economy of time in doing the daily chores is well illustrate, showing the amount of travel in doing certain chores when the buildings are poorly located and the saving made by moving and relocating some of them.

Building Plans

The plans of the buildings should be considered quite as much as the plans for the farmstead. Before erecting a building, a plan should be made of it, drawing to scale, with all of the interior arrangements made. Feed storage rooms should be provided which will permit quick and easy feeding of the live stock. Convenient arrangements should be made for watering, and an attempt made to lessen the labour in every way possible. The harness rooms, milk rooms, and grain bins should all be at convenient points and located with the view of lessening the amount of travel. If a harness room is provided, it should be at a central spot and where every stall can be reached with the minimum amount of travel.

The milk room, where included in the barn, should be located near the centre but where it can be well ventilated and where the objectionable odors of the barn can be kept from it. The grain bin frequently can be placed overhead, with an elevator run by a gasoline engine for elevating the grain into it. By providing suitable spouts, feed boxes, and alleys, the work of feeding the animals is thus greatly reduced. Attention to details of arrangement is the first essential in providing economical and convenient buildings.

6

Integrated Fish Farming

INTRODUCTION OF INTEGRATED FISH FARMING

Chinese integrated fish farming system is so broad in scale with so many models that there is no comparative in the world, and it has developed its own characteristics as well. This technology attracts the world's attention. Perhaps, the reason for this is that it can fully develop and utilize local natural resources and it can produce more materials for the people and it can satisfy the needs of the world economic development on many aspects.

THE ADVANTAGES OF INTEGRATED FISH FARMING

Establishment of a man-made ecosystem without any wastes

In developed countries, the more their industry develops, the more intensive their farming is, and the larger the scale of chicken and pig raising factory farms is, the more the wastes accumulate, e.g. in Japan, only the excreta of cattle, pig and chicken reach more than 70 million metric tons per annum. If this great amount of excreta is not disposed of, it must pollute the environment.

The annual output of animal manures is considerably high —10.4–11.8 tons per black-and-white cow; 4.1–4.7 tons per pig; 0.04–0.07 tons per duck. The decomposition of so much excreta produces gaseous ammonia, etc. The odour pervades everywhere. As a result, the water, the land and the air are all polluted and it will jeopardize the people's health.

In fact, the manures of livestock and poultry are good organic fertilizers for fish farming. About 40–50 kg of organic manures can be converted into one kg of fresh fish. If the farm combines fish farming with mulberry cultivation, sericulture and silk extraction from the cocoons, the pupae will be a fine feed for fish while the worm faeces and waste water obtained at the processing factory are used as pond fertilizers. In an integrated fish farm, the manures of livestock and poultry and the dregs, lees and waste water obtained from starch processing and wine brewing are all used for fish farming. The pond silt can be used as fertilizers for terrestrial fodder crops which can in turn be used to raise livestock and poultry or directly for fish farming. Thus, it forms a recycling ecosystem, which utilizes various wastes and afterwards, an agricultural ecosystem can be set up without any wastes.

Increasing the food supply for the mankind

Short supply of food and protein is a serious problem that we are now facing. Therefore, merely using the pelleted feeds of grains and animal protein as fish feeds is not economical and often makes the food for the mankind more insufficient. If the grains such as wheat are used in fish culture, the output could be up to 4815–9750 kg/ha. But the average Food Conversion Rate of grain is 3, i.e. 3 kg of grains in dry weight could converted into 1 kg of fresh fish. This kind of fish culture is not suitable for the countries which want for food. However, the natural food organisms cultured in fish ponds by using organic manures could totally take the place of pelleted feeds or grains.

The quality of fish will not be changed by using animal manures as fertilizers in fish culture. The daily output could reach 15–32.25 kg/ha or even higher. E.g. Helei Fish Farm in Wuxi has not only built fish ponds with an area of 69.4 ha but also a dairy with 100 cows and pigsties with 1000 pigs and duck yards producing 10,000 eggs per day. It has become a subsidiary food production base, which is an integration of aquaculture industry and commerce. So far as aquaculture is concerned, they combine fish farming with livestock and poultry raising, however, putting fish farming first. In 1981, it supplied not only large amount of fertilizers to produce 600,000 kg of fish for market but also supplied 490,000 kg of pork, etc.

The scope of integration in an integrated fish farm could be considerably wide. Apart from fish cultured in the water body, the water surface can be used for goose and duck raising, pond dikes for fruit tree and mulberry cultivation or for setting up pig-sties, slopes for fodder crops. The products from an integrated fish farm are not only fish but also meat, milk, eggs, fruit, vegetables, etc. It is apparent that integrated fish farming can fully utilize the water body, the water surface, the land and the pond silt, etc. to increase the food supply for people.

More job offer

In unitary fish farm, the labour force cannot be reasonably employed while the integrated fish farm may offer more jobs, for example, in Helei Fish Farm, the occupations except fish culture offer 149 jobs, among which there are 48 persons for duck raising, 19 for cow farming, 14 for pig raising.

Reducing the cost, raising the output and increasing economic benefits

Nowadays, the problem that the aquaculture in developed countries is facing is that the cost of pelleted feeds is too high. It is related to energy crisis and protein shortage also. The integrated fish farm produces feeds and fertilizers for itself, savrer energy and reducing. expenditure considerably, e.g. in 1981, Helei Fish Farm produced 5.5 million kg of pig manure, 1.85 million kg of cow dung, 1 million kg of duck manure, 9.5 million kg of waste water from silk extracting workshop. The total amount of organic fertilizers was 17.65 million kg. Outflow of animal excreta by gravity could reduce the production cost by about 97 year each mu of fish pond. In 1976, Helei Fish Farm began to put duck, cow and pig raising into integrated fish farming system, food processing industry in 1979 and commerce in 1980. During the period of 1977–1981, the yield of fresh fish increased from 115,000 kg to 600,000 kg (the same area) while the yields of livestock and poultry increased from 135,000 kg to 490,000 kg. In 1981, average income per capita increased by 120. over the year of the establishment of the fish farm in 1966. The annual net fish production of integrated fish farms, only by using organic manures.

THE CHARACTERISTICS OF CHINESE INTEGRATED FISH FARMING

The integrated fish farming practices have developed into a complicated structural network in line with the local conditions.

China is a vast country with a large population and varied natural environments. The agricultural structure and the economic conditions of each locality are also different. Therefore, various integrated fish farming systems in line with the local conditions have developed into a complicated structural network. The Pearl River Delta is located to the south of the Tropic of Cancer. Its annual solar irradiation is 110 kcal/cm^2 but averaging 60 Kcal/cm^2. Average temperature is about 22°C; 2–3 days of frost a year; more rain and high temperature in summer with maximum temperature of about 37°C and relative humidity 76–85%; annual sunshine time between 2000 and 2500 hours. Such geographic and climatic conditions are very conducive to the cultivation of mulberry trees, sericulture and fish farming. Hence, the farmers in the Pearl River Delta through production practices have been able to take advantage of local natural resources by integrating mulberry cultivation and sericulture with fish farming, which leads to the establishment of a complete, scientific "mulberry plot-fish pond" man-made ecosystem.

Even in the same geographical zone, the items of integration by various production units are different. Helei Fish Farm, Xinan Fish Farm, Helei First Fishery Brigade, Wangzhuang Fish Farm as well as the Municipal Fish Culture Farm in Wuxi practise different items of integration. Liutan village has recently incorporated agriculture, sideline occupations, aquaculture and commerce into its integrated system. The integrated fish farm was established in 1980. The farm site was earlier a water-logged paddy field of 17.33 ha, subject to annual flooding and with low yield of agricultural crops. Based on the topography of the low land, the farmers restructured the land into fish ponds of 10.67 ha. in the winter of 1979 and the spring of 1980.

The farm consists of 15 grow-out ponds (8 ha), 11 fingerling ponds (2.67 ha). Apart from pig raising, green fodder planting is the major item of the integrated management. It has 1.33 ha of pond dykes for the cultivation of English rye grass (Lolium pereme), sow thistle (Lactuca indica), etc. They utilized water cane shoots field to cultivate duckweed with average yield of 37,500–45,000 kg/ha. With sufficient supply of green fodder, Grass carp and Wuchang fish (Megalobrama spp.) become the dominant cultured species, with the yield of 5250 kg/ha, that is 45% of the total pond fish production. It indicates that the maximal efficiency of the integrated fish farming system can be obtained only through maximal utilization of natural conditions and the agricultural characteristics of the regions concerned.

The socio-economic conditions should be taken into consideration in developing integrated fish farming. The fact is that the diversified economy develops in the harmonious interaction among the socioeconomic conditions, agricultural production, and natural conditions on the basis of regional climatical differences.

Since the development of integrated fish farming is site-specific and the conditions of each site are different and complicated either from the macro or micro view points, the linkage between the trades of integrated fish farms in China become a complex matrix. Simple modele involving mono-integration of fish-cum-animal husbandry, fish-cum-poultry or fish-cum-crops are now getting fewer and fewer in China with the exception of small-scale individual farms.

Utilising wastes by various ways of recycling

The integrated system of fish farming is rather complicates involving various forms of integrations and methods in utilization of organic wastes.

The animal excreta alone can be effectively utilized through employing various methods and techniques.

1. Fresh animal manure can be applied directly to the fish ponds. pigsties, poultry coops and pens for ducks and geese can be constructed on the dikes or above the ponds. Fresh manure thus enter the ponds directly, avoiding energy losses due to processing of manure and transportation. The feedstuff of livestock are not fully digested and then can be directly utilized by fish. Therefore, the number of animals should be compatible with unit water surface.
2. The residues and liquid part after anaerobic fermentation to produce biogas are used for fertilizing fish ponds.

   Compost after aerobic fermentation can also be used for fish culture.
3. Animal manure can be used indirectly through one or two trophic levels in a food chain such as growing fodder crop to feed herbivorous fish, producing earthworms or other animal-feedstuff for carnivorous fish directly or used as part of the composition of pelleted feeds.
4. Poultry manure can be used to feed pigs and pig manure can in turn be used as fertilizers for fish pond.

FURTHER DEVELOPMENT AND RESEARCH NEEDS

The prosperity of agricultural economy and the application of new technology stimulated and improved the production of integrated fish farming, e.g. the use of aerators and the mechanization, to certain extent, raised production efficiency of the classical mode of Chinese fish culture. Although integrated fish farming has long been practised in China and rich experience has been gained through years of improvement and considerably high yields have been achieved, the scientific basis for some of the existing techniques are still awaiting clarification.

In order to raise the efficiency of integrated fish farming considerably, the biological basis of integrated fish farming must be studied, finding out the rule and developing aquaculture technology. In theory, two ecosystems must be clarified: First, the integrated fish-livestock-crop land-and-water ecosytem research should be carried out on its structure and functions so as to set up an optimal and harmoneous ecosystem on a perfect structure and with sound functions; second, the pond ecosystem.

The structural research concerns the biological interaction among fish, livestock and crops and their matched proportion. Material cycle and energy flow should be measured to elucidate the biological relationship and exact proportion of quantity among these production links. So far as the land-and-water structural ecosystem of integration is concerned, crops are producers; livestock and fish are consumers; aquatic organisms and soil organisms are decomposers.

It is a rather complete ecosystem. In this system, benign circulation demands that there be no wastes in any link of this structure and the production be not hindered by lack of sufficient energy. For this reason, it needs a thorough research on fish pond ecosystem. Manureloaded pond is a semi-closed man-made ecosystem. Being put into fish ponds, animal manures begin to enter into the process of decomposition by bacteria. It's a complex process of pond dynamics in which animal manures will be converted into fish protein. It involves many factors such as physical or chemical parameters or biological factors in food chains.

In order to investigate the rule of variation and intricate relationship between biology and non-biology, the research needs cooperation and common efforts of different scientists and specialists from multidisciplines such as aquaculture, ecology, botany, microbiology and chemistry.

Integrated fish farming is a low energy consumptive but high efficient aquaculture system. It's a right way to develop fresh water fish culture and also it's a strategic measure of developing Chinese aquaculture. Integrated fish farming will play a promising role in solving the problem of "hard to get fish".

FISH-CUM-PIG INTEGRATION

Fish culture combined with pig raising is a traditional integrated fish farming model in China. Now, a few of other countries are also engaged in fish-cum-pig integration. Since 1974, Dr. Buck has conducted experiments on utilization of pig manure in fish farming. In 1977, he polycultured Cyprinidae in the test pond with Silver carp as major species. The output was 4585 kg/ha.

From economic view point, pig-raising can be divided into three situations: loss, balance and profit. If it combines with duck raising and fish farming, not only the economic efficiency could be increased but also the social efficiency and ecological efficiency could be raised. The pig food are often leftover and residues from the kitchen, aquatic plants and agricultural products and their wastes. The pig excreta is in turn used as organic manure in fish ponds. It might keep the environment clean. Pork is the indespensable subsidiary food. Pig manure is of high quality by virtue of nutritional completeness. That's why farmers usually combine fish farming with pig raising and it becomes a common pattern of integration in China's rural area.

Methods of pig manure application

There are two types of pig sty in China. One is the simple pig shed constructed on the pond dyke or over the water surface; the other is centralized hog house. Both types have their own merits and demerits. The cost of the simple one is lower and moreover, it's easy to apply the manure to fish ponds. Therefore, this type is more suitable to household or small-scaled farms of fish-cum-pig integration. The excreta of pig can automatically flow or be flushed into the pond. It can save much labour. If the area of a fish pond is less than 8 mu, a pig sty can be set up on the pond dyke and then, pig wastes after flowing into the pond, can diffuse to the whole pond by virtue of winds and waves. If more than 30 pigs are raised in the same spot, the method of flowing by gravity is not suitable because the more the pigs, the more the pig manure. More often than not the place near the pigsty will be heaped up with manure and the water quality will be partially deteriorated. If pig manure sinks to the bottom too much or flows into the pond too much through the pipe, it will also cause fish surfacing.

In order to fully utilize the effectiveness of manure, it's necessary to pay attention to the method of applying. The manure from the centralized hog house is easy to be concentrated to a storage pond or a sedimentation basin. The amount of application can be controlled by various means. Centralized hog house is fit to be built in large-scaled integrated fish farms.

The manure after dilution can be spread along the pond dyke by manual labour or by small boat in small fish ponds. If the fish pond is large, it's better to use boat and mechanic apparatus so as to spread the manure evenly. Here are two methods:

To use stern-thruster boat (outboard motor boat) for fertilization Hang the iron cage loaded with manures alongside the boat. The space between grids is about 2.0—2.5 cm. The capacity of the cage differs with the amount of fertilizers needed. The cage loaded with manure must be hanged 10—20 cm below the surface. When the engine starts, the current formed by the thruster will rush against the manure in the cage and as a result, it can be spread evenly.

To install a pump in a boat

The manure should be diluted in a cabin through a funnel and then be spread into the whole pond.

The large quantity of liquid manure in sedimentation basin of largescaled integrated fish farm can be introduced through a pipe or a hose to the pond side and then be sprayed through a nozzle. The average rate is 200 litres/min each nozzle. The nozzle should be fixed 0.5—1.0 m above the surface, in this way, the liquid manure can be fully contacted with air. The nozzle will serve as an aerator if it works to spray water when the D.O.C. in pond water is low.

In fish-cum-pig integration, two points must be especially noticed: one is lack of oxygen in the pond; the other is oligotrophic pond water. That's why the water quality should be monitored at any time. Besides, the production period of pig should match the demand of pig manure in fish farming.

Integrated fish farms are chiefly engaged in raising hogs. The advantage is that the production periods of pig and fish farming match well. Hog can be raised for 2 periods a year. The duration of each period lasts for 5 to 6 months. 60% of the total amount of manures applied to fish ponds are given during the first half of a year. The peak of application is in Jan. and Feb. for base manure and in June and July for the additive manure. Usually, no manure is applied after the middle 10 days of Oct. Therefore, one pig production period is supposed to be from the middle 10 days of Feb. to the middle 10 days of Aug; the other is from the middle 10 days of July to the middle 10 days of Jan. The growth of the first batch of pigs peaks at the latter part of the production period and their excreta as well. It can just meet the fertilizer demand of fish farming. The excreta of pigs from Nov. to Jan. serve as base manure for fish farming next year. This collocation not only can meet the demand of pig manure during the whole period of fish farming but also satisfy the concentrated demand of pig manure in fish farming.

RATE OF PIG RAISED PER UNIT AREA OF FISH PONDS

The actual rate of pig raised for per unit area of fish pond in integrated fish farms in different areas is from one to five per mu of fish pond. How much excreta of pigs can the given unit area of fish pond accept so as not to deteriorate the water quality of the pond but to maintain sufficient food for fish? It must be worked out through the productive experiments in line with the local conditions. In China, a fish pond is often given various kinds of organic manures such as pig manure, cow dung or other animal manures so the quantity of applying pig manure must be reduced correspondingly.

The capacity of a fish pond to accept animal manures depends upon the techniques of application and the nature of manures. In Hungary, the growth period of fish is 150—180 days. If pig manure is applied by the method of stacking in a corner, the rate of application is 100—134 kg/mu of fish pond. If the method of so-called carbon manuring technique is adopted, that is to say, fresh pig manure mixed with pond water is spread to the whole pond frequently, the rate of application is 20—40 kg/mu/day. It's an equivalent to 1000–1500 kg of condensed liquid manure or 1.2—2.5m^3 of solid wastes of commercial pigsty.

In theory, the maximum capacity per mu to accept manures is 2 to 3 times the above mentioned figures. Why, it involves many facets such as environmental conditions of a fish pond, the quality of pig manure, the managerial techniques of fish ponds and so on.

Fish species cultured in fish-cum-pig integration

The production efficiency of fish-cum-pig integration appears in the full utilization of food organisms by fish in pig manured pond. In 1950's, when Hungarian began to culture fish by using pig manure, two problems happened. (1) Water bloom of Aphanizomenon flosaquae and other blue algae appeared time and again because of non-control of manure application. It causes the disappearance of Daphnia spp.

Common carp can not utilize plankton directly, resulting in waste of primary productivity, The filtering species should be stocked in manured pond, and thus, it will avoid auto shading of phytoplankton and decreasing the products of photosysnthesis. For this reason, Hungarian have stocked certain kinds of Chinese carps since 1970. Silver carp and Bighead are the best species to control plankton. They have fine filtering organs and can filter and eat phytoplankton in size above 220ì. If the water temperature goes up to 22°C, they can filter and eat more plankton and a great number of detritus and bacterial congregations in a given unit time. The feed basis formed after fertilization could be utilized more. In mono-pig-manured pond, a few Grass carp may be stocked to control Aphanizomenon flosaqae and submerged vegetation in the pond. In mono-culture of Common carp, 50 kg of pig manure can be converted into 1.25–1.5 kg of Common carp; in polyculture with Common carp as major species, 50 kg of pig manure can be converted into 1.75–2 kg of fish; in polyculture with Silver carp as major species, 50 kg of pig manure can be converted into 3 kg of fish. The feeding habit of Tilapia is wide in scope and serves as cleaner in pig manured pond. This species fits the pigmanured pond. In Hubei Province, China, Silver carp, Bighead and Tilapia are the main species in pig manured polyculture, assorted with a few Grass carp, Common carp and Xeno cyprinus as minor species.

Economic efficiency of fish-cum-pig integration

The economic efficiency is apparent if the wastes of pig raising can be utilized. In Hunan Province, one mu of fish pond matches with three pigs. In polyculture with Silver carp and Bighead as major species, the output could reach 150–200 kg/mu. In tropic and sub-tropic zone, the temperature is higher and the biological process becomes quicker, more pig manure can be decomposed in the pond and moreover, with proper species in polyculture, the utilization rate of pig manure will perhaps be much higher.

Fish-cum-pig integration can reduce the production cost. For example, in 1981. Xinan Fish Farm Wuxi gained 3388000 kg of pig manure including 50% flush water. It saved 14229.6 yuan of manure expenditure. The fish yield was 33880 kg. The profit of pig raising itself was not too much, nevertheless, the total revenues of integrated fish farm increased because the excreta of pig took the place of artificial feeds and inorganic fertilizers, which accounted for 58.8% of the cost of fish farming.

However, the simple fish-cum-pig integration becomes less in China. This pattern is usually a part of fish-livestock-agriculture integration system which is a comlete man-made ecosystem.

Fish-Cum-Cow Integration

Fish farming using cow manure has long been practiced in China. Since the founding of the People's Republic of China, cow farming has developed rapidly. It promotes fish-cum-cow integration, which is one of the common modellings of integration in the southern part of China. On fish farms cow farming can save much fertilizers, cut down fish feeds and increase the income from milk while fish farming on cow farms can make it realized that "the excreta of a cow can be converted into a thousand jin of fish". Moreover, cow manure can be disposed at hand resulting in saving much money, labour and energy and also it is good for the environment. If farmers use part time worker to conduct small scale integrated fish farming with cow raising, they can not only earn more money but also can supply both fish and milk to the market.

THE MATERIAL BASIS OF FISH-CUM-COW INTEGRATION

In fish-cum-cow integration, cows are suppliers of the materials for fish farming. They can provide cow manure, leftover and matted grass of cow sheds, etc.

The biological basis of fish farming using cow manure

Among all the excreta of livestock, the amount of excreta of cows is the greatest and the most stable. If a cow is 460 kg in body weight, it can produce 0.081 kg of manure/kg of body weight per day and the annual amount of faeces can reach 13600 kg and that of urine about 9000 kg.

The nutritive content of cow dung is a little less than that of pig faeces. If 0.024 kg of fresh cow manure is applied to a water body of 1 cubic meter per day, inorganic N and P in fish ponds will be 0.897 mg/L and 0.024 mg/L respectively, which are close to the inorganic N, 0.97—2.06 mg/L (average 1.38±0.42 mg/L) and the inorganic P, 0.018—0.036 mg/L in high-yield fish ponds. The ratio of N/P will be 36.9, which is a little higher than the value of the ratio of N/P in phytoplankton. Apparently it will limit the large increment of phytoplankton in fish ponds. Nevertheless, the average amount of phytoplankton in manured pond can still reach 19.15±6.5 mg/L which is apparently higher than that in control ponds and close to the lower limit of optimum food density (20—100 mg/L) of Silver carp and Bighead. The average biomass of zooplankton will amount to 5.61 mg/L which is much higher than that in unmanured pond and the ratio between the biomass of zooplankton and phytoplankton is 1/3.4.

Both the biomass of zooplankton and the ratio between the biomass of zooplankton and that of phytoplankton come to the level as much as in fertile water fish pond (that is to say, the biomass of zooplankton is between 5 and 30 mg/L and the ratio is between 1/4 and 1/3).

No risk in fish farming using cow manure

Cows are ruminants. It is due to repeated grinding and digestive decomposition catalyzed by great amount of micro-organisms in rumina that cow manure is very fine. It can suspend longer in the water. The sinking speed of cow manure particles is 2.6 cm/min and that of pig manure particles 4.3 cm/min. If the same amount is applied, the sediment of pig manure is 33% higher than that of cow manure after 24 hours while the suspended organic detritus below 0.65u in size in cow-manured pond is 39.99 mg/L, which is 153.4% higher than that in pig-manured pond. And it occupies 54.6% in total amount of suspended particles, which is the highest percentage compared with that in pig, duck, or chicken-manured ponds.

The suspensibility not only enables fish to get more feeds but also reduces oxygen consumption caused by manure stacking and avoids forming of harmful gases. The BOD of cow manure is lower than that of other livestock manures because the cow forage has already been decomposed by micro-organisms in cow's body. The BOD of 1 kg of cow manure in five days is 20.6g, 32% lower than that of pig manure which is 30.0g. The same tendency appears when they are in fish ponds.

In one week's consecutive measurement, the lowest dissolved oxygen content in cow-manured pond at 5:30 a.m. is 1.8 mg/L while that in pig-manured pond is 1.0 mg/L. The average dissolved oxygen content is 4 mg/L in cowmanured pond but only 3.3 mg/L in pig-manured pond. By virtue of suspensibility and low oxygen consumption of cow manure, fish farming using cow manure is more safe. Freshwater Fisheries Research Center conducted an experiment on different livestock manures in fish farming in 1983. It indicated that the survival rate of fish in cow-manured pond is 98%, second to none.

Wasted food of cows for fish farming

Cows feed chiefly on grass. During the grass growing season of about 7 months, the fresh fodder grass for each adult cow amount to 9000—11000 kg. But the leftover occupies 27.9%, around 3000 kg. However, that period of time happens to be the highest ingestion seasons of herbivorous fish, therefore, the wasted fodder of cow can be utilized as fish feeds. The results differ with different quality of grass. The food conversion coefficiency of terrestrial wild grass is about 40–50. The matted grass in cow shed can only be used as compost of the pond but the leftover of fine fodder for cows can be used as fish feeds.

MANAGEMENT OF FISH-CUM-COW INTEGRATION

Disposition and proportion

In the integrated fish farm, cow sheds should be built in the vicinity of fish ponds so as to utilize cow manure at hand. The faeces and urine are collected separately, the former transported by convyor or boat or car and spread evenly, the latter pumped into fish pond. If the floor of cow shed is higher than the pond dike, manuring ditch could be dug to collect faeces and urine together, then to flush them into fish ponds. It saves time and labour. But the above methods often cause stacking of manure and uneven fertilization. The best way is to mix up cow dung and urine and to pump them into fish ponds evenly. If the hose pipe has a spray nozzle, the result will be better. If the area of fish ponds is large, the boat dragging a stern manure bucket could be used. Then manure will be evenly spread by the waves caused by sailing.

Proportion here means the area of fish pond matched with one milk cow. It depends on synergism of many factors, such as the amount of cow manure and wasted food; the species ratio and target output of fish; the sediment of pond; the quality and quantity of pond humus; other manures and feeds applied. Supposing fish culture wholly depends upon cow manure and wasted food of cows with a net production target of 250 kg, among which 10% is herbivorous fish, the rest filtering and omnivorous fish, each milk cow can provide manure and feeds for fish in 2 mu of fish pond. If the proportion of feed-eater increases, the method of calculation of fish-cow proportion is shown in the empirical equation related to manure demand and quantity of livestock and poultry in model plan.

Period and frequency of manuring stands on the transforming speed of cow manure and season change and the rule of the ebb and flow of natural food organisms in fish pond. In 1983, Freshwater Fisheries Research Center carried out an experiment on measuring the peak time of N and natural food organisms at different water temperature in a fish pond manured once a week. It showed all nutritive factors had two peaks in a week. The high peak appeared earlier when the water temperature rose. Therefore, the guidelines to determine period and frequency of manuring are as follows: In late winter and early spring, with sufficient base manure, a few times but in large quantities, that is once every 5–7 days; in spring and autumn, once every three days; in summer, once every day or every two days in small quantities and it's better to manure everyday and make some adjustment according to weather, water colour and fish growth.

Stocking pattern in fish farming by applying the waste in cow raising Cow dung and urine is beneficial to filtering and omnivorous fish culture. Therefore, Silver carp and Bighead are taken as major species assorted with omnivorous fish (Common carp as main minor species), and 15—20 % of herbivorous fish. If it's over that ratio, supplemented feeds must be applied. It's optimal that the output of herbivores occupies around 12% of total net output of the pond.

ECONOMIC BENEFITS OF FISH-CUM-COW INTEGRATION

Both milk and fish harvests

Each black-and-white milk cow after three births can produce around 5000 kg of milk annually, which contains 155 kg of protein. Except the milk for calves, each cow can not only supply the market with 4500 kg of milk, but also can provide feeds and manures for fish farming, which in turn can produce 500 kg of fish containing 55.8 kg of protein annually. Each cow can provide a total of 210.8 kg of protein that is 3700,000 kcal in calorie value annually. Fish-cum-cow integration can increase protein by 36 % and calorie value by 10.6 % above the unitary cow farming and it can increase protein by 2.7 times and calorie value by 9.5 times over those produced by unitary fish culture.

Increasing revenues, decreasing expenditure

Fish-cum-cow integration can get more money from fish, milk and calves. The output value of each milk cow is averagely more than 3100 yuan. The excrements and leftover of one cow can be converted into 500 kg of fish without additional feeds and manure. Therefore, it can reduce half of the cost of fish farming.

Alleviating unemployment

In fish farming, raising four cows can offer one opportunity of employment for people.

Saving energy

Fish-cum-cow integration can provide feeds and manure for fish farming so that it can save a lot of energy for transportation. However, the investment of fish-cum-cow integration is much greater at one stroke. It needs about 2500 yuan RMB for each milk cow and attached facilities. On the basis of investigation, the profit rate of cow is higher, which is about 30%, that is to say, 100 yuan net profit can be gained from 300 yuan fixed capital. In term of net profits and cost saving in fish farming, the economic return of investment will cover less than three years.

MODELS OF FISH-LIVESTOCK-CROPS INTEGRATION

The model of fish-livestock-crop integration is the transverse combination of fish-cum-livestock and fish-cum-crop models, namely, to utilize animal manures and pond silt from fish-cum-livestock integration as agriculture manures and to utilize crop products from fish-cum-crop integration as feedstuff for fish and livestock. So far the most popular model is pig-grass (vegetables)-fish and vegetables-pig-fish integration, etc. The former develops more rapidly nowadays.

In fish-cum-pig integration, all pig excreta are used to fertilize fish ponds with plankton-eater as major species whereas in pig-grassfish integration, part or the whole of pig excreta are used as manure for terrestrial and aquatic fodder plants with high-yielding pasture grasses as major crops. Green fodder crops are used to culture herbivorous fish and bring along rearing filter-feeding fish. No or just a little pig excreta go down directly to fish ponds. Finally, pond silt accumulated in fish farming returns to the land as crop fertilizer. The new nutrient cycles and energy flow is thus set up in pig-grass-fish integration. Both terrestrial and aquatic primary productivity is fully utilized.

Therefore, its structure is more rational and more effective in comparison with fish-cum-pig integration. It can make up for insufficient manures in fish-cum-grass integration.

ECOLOGICAL EFFICIENCY OF PIG-GRASS-FISH INTEGRATION

To raise utilization efficiency of light energy

The products of photosynthesis depends on the amount of light that is the product of light intensity by sunshine time. Terrestrial pasture grasses can absorb light energy from the sun directly and can avoid shading each other under artificial control, e.g. mowing in time, so as to reduce the reduction rate of radiant energy.

The rate of reduction of radiant energy in water is higher than that on land because the water surface reflects and scatters light. When the sun rises high, the light intensity will be reduced 6% because of reflection of calm surface or 10% because of reflection of waving surface. When the sun rises or sets, most of light can not enter into water due to the shading of pond sykes, therefore, so far as the sun light is concerned, daylight in water is shorter than daylight on the land.

If the gradient of a pond dyke is 30 degree the sunshine time of the pond water along the pond dyke is reduced by one third in comparison with that on top of pond dykes.

The lower the water level and the smaller the pond area, the shorter the sunshine time. Since there is a large amount of suspended and dissolved substances in fish ponds and especially plankton propagate greatly, auto-shading occurs and it's hard to control. According to the experiments of Freshwater Fisheries Research Centre, the transparency and the amount of suspended substances are in inverse proportion. In a natural pond, the coefficient of correlativity r=-0.9083. The regression equation is as follows:

$Y = 247.692 - 2.840\ x$

Y = transparency (cm)

X = the amount of suspended substances (mg/L)

Larcher (1980) realized that radiation intensity is in inverse proportion to the water depth, decreasing in the form of exponent (or index). When the sky is evenly clouded, the light intensity, according to synchronous measurements, on the soil surface with the depth of ryegrass of 40 cm, is 900 lx, which is 13.43% of that on the surface of grasses and that means its relative light intensity reaches 1.8% of 50 klx, the saturated point of net photosynthesis of land grasses; the light intensity at 40 cm below the water surface of a pond is 28 lx, the relative light intensity is only 0.42% of that on water surface. It only reaches 0.14% of 20 klx, the saturated point of net photosynthesis of phytoplankton. Moreover, 40 cm is the maximum height of ryegrasses to be mowed. In the growing or ratooning phase before mowing, the relative light intensity is much higher than the measured one while the relative light intensity in the pond cannot be increased artificially. Therefore, from the view point of light intensity and sunshine time net photosynthesis of phytoplankton cannot compare with that of terrestrial plants either in a specific time or in a period of time.

Terrestrial grasses have greater ability to photosynthesize than phytoplankton. Under the normal supply of carbon dioxide and with an optimal light, temperature and water conditions, the uptake of carbon dioxide of sunplant herb is 20–50 $mg/dm^2/hr$. The intake of energy and the primary productivity of terrestrial grasses are higher than those of phytoplankton. Manured with pond silt alone, the yield of ryegrass can reach 4500 kg/mu in Apr. and May. The average daily output is 82 kg in May. The average daily yield of Sudan grass can reach more than 100 kg/mu.

The net output of oxygen produced by phytoplankton in Wuxi or Xiamen is 8–10g o_2/m^2/day. So the net yield of phytoplankton is only 42.3–52.9/ma/day. (Le xing-zhi et al). From the measurement in Wuxi high yielding fish pond, the net yield of phytoplankton in peak period was 41 kg in wet weight per mu per day, only 50% of the net yield of rye grass, 41% of the net yield of Sudan grass per mu per day.

After calculation, in accordance with the balanced equation of photosynthesis, 1 kg of O_2 is equal to 6.1 kg of phytoplankton in wet weight while 1 kg of O_2 means producing 3.51 Mcal of energy. Therefore, 1 kg of phytoplankton averagely holds 0.575 Mcal of energy. Daily output of plankton from May to Aug. is 41 kg/mu, that is, 23.575 Mcal. 1 kg of ryegrass holds 1.035 Mcal. Daily output of ryegrass in May holds 84.87 Mcal, that is 3.6 times that of phytoplankton. 1 kg of Sudan grass holds 0.75 Mcal. Daily output from June to Sept. is 77.25 Mcal, which is 3.3 times that of phytoplankton.

In the case of rye grass and Sudan grass. We can see that the primary productivity of terrectrial grasses is 2 to 4 times that of phytoplankton in high yielding fish ponds and the output of energy 3.3–3.6 times. The yield of terrestrial grasses can be further increased through cultivation and popularization of good varieties of seeds, rotation systems and intercropping with leguminous grasses and scientific management. On the contrary, how to increase the primary productivity of phytoplankton is a hard nut to crack so far.

The utilization rates of primary productivity of phytoplankton and terrestrial grasses by fish are different. Furthermore, phytoplankton can continue their propagation only after part of seeds remain in a pond. He Zhi-hui (1983) calculated the consumption of phytoplankton in high yielding fish ponds in Wuxi. The direct phytoplankton consumption of Silver carp, Bighead, Tilapia and Crucian carp or through zoopalnkton accounted for 82% of the biomass of phytoplankton. (70% of the zooplankton are consumed by plankton-eater.); whereas the utilization rate of ryegrass with the height of 50 cm or Sudan grass below 1.2 meter by fish is 100%.

To utilize two primary productivity

Compared with manuring the pond water, pig-grass- fish integration utilizes high effective primary productivity of terrestrial grasses and at the same time, it also uses the primary productivity in fish ponds. In this system, herbivorous fish are the major species, which are stocked according to the stocking pattern with net yield of 400 kg/mu. The mean capacity load of herbivorous spp. is 190 kg. The daily Sudan grass ingestion of herbivorous fish occupies 22.5–30% of the body weight of fish, average 25%.

Thus, the ingestion of herbivorous fish is 48 kg/mu/day. The excreta of herbivorous fish are about half of its ingestion amount, so the daily amount of excreta are 24 kg. On the basis of analysis of South-Western Normal College in China (1977), the N and P contents of Grass carp excreta are 1.102% and 0.426% which are three times and two times those of pig excreta respectively. At least, the daily excreta of herbivorous fish are equal to 24 kg of net pig manure. During the 250 days of ingestion period, the amount of excreta of herbivorous fish are equal to 6000 kg of net pig manure or the amount of net excreta of 6 pigs. It's near to the total amount of manure application in the fish pond with Silver carp and Bighead as major spp.

Thus, growing grass by using pig manure can provide sufficient fodder grasses to the herbivorous fish, the excreta of which, in turn, can supply nutritive salts enough for the fish pond so that the primary productivity of the fish pond can reach the mean level of the pig-manured pond. In Wuxian, Zhangzhuan Aquaculture Brigade which has adopted pig-grass-fish integration, all the pig manure are used to grow grasses. No additional manure is applied to the fish pond with herbivorous fish as major spp., which occupy 55–72% of the total stocking amount. The water colour looks green and brownish green with transparency of 10–35 cm. For example, in pond No. 33 with herbivorous fish occupying 72%, the average transparency is 17.5 cm. According to the regression equation of phytoplankton and transparency after He zhi-hui.

$$Y = 153.69 - 3.06x$$

Y = the biomass of phytoplankton (mg/l)

X = transparency (cm)

The biomass, of phytoplankton of that pond is about 100 mg/l. The biomass of phytoplankton of high yielding fish ponds in Wuxi is between 20 and 100 mg/l. It indicates that the excreta of herbivorous fish can fertilize the pond water and provide sufficient biomass of plankton demanded by plankton-eater.

Pig-grass-fish integration can avoid certain ecological defects caused by direct application of pig manure in ponds. The direct application increases the BOD and lowers the dissolved oxygen levels esp. if stacking application of manure is adopted at ten-day intervals. The BOD_5 = 29.09 – 30.75 g/kg. According to the experiments of Fresh water Fisheries Research Centre, the average BOD_1 of fish ponds are 4.10 mg/l and 5.48/1 respectively when pig manure is spread evenly into the pond at the rate of 0.12 kg and 0.29 kg/m^3/ week.

At 5:30 in the morning, the average D.O. is only 1.0 mg/l. Zhangzhuan Aquaculture Brigade adopted pig-grass-fish integration in 10 ponds, and supplement molluscs and fine feeds without any other manures in fish ponds. The carrying capacity of 9 ponds were from 779.3 to 1350.5 kg/mu except one pond with 475.3 kg/mu. All Dissolved oxygen levels were high. The minimum DOL was 2 mg/l and average 4.6 mg/l in Pond No. 10 with the highest carrying capacity.

Because this system utilizes two primary productivity, avoiding some ecological defects, there is a greater potentiality of increasing fish yields. In terms of many reports from abroad, the fish yields of filterfeeding species and omnivorous species in a manured pond only reach 15–31 kg/ha/day when they are monocultured or polycultured as major species. Generally speaking, the yield of plankton-eater is 300–400 kg/mu. It's difficult to surpass this figures and moreover, the economic efficiency will be reduced when the yield increases. Nowadays, in most of fish ponds with the yield of 500 kg, the herbivorous fish are cultured as major species with grass as main food and without application of manure or with just a little manure.

The excreta of 1 kg of herbivorous fish can fertilize the pond water and produce 0.2–0.5 kg of Silver carp and Bighead. The larger the size of fingerlings the more their ingestion, and the more the number of planktoneater that they bring along. In Grass carp rearing pond with gross yield of 110–150 kg compared with Grass carp rearing pond with gross yield of 50–100 kg, the net yield of plankton eater is 37.5 kg higher.

In fish-pig link, filter-feeding fish and omnivorous fish feed on detritus and the bacteria on detritus. The relationship between the fish yield and both detritus and bacteria depend on their quality rather than on their quantity, therefore, it can not be overestimated just according to the amount of detritus in a water body. The experiment of Lin Wan-lian also proves on one aspect that it is a loss to use pig manure to grow grasses for fish farming so far as part of detritus, which can be utilized by fish, is concerned. Nevertheless, the good effects of pig-grass-fish integration can make up for the loss. What's more, the effects of detritus is not so good.

THE ECONOMIC EFFICIENCY OF PIG-GRASS-FISH INTEGRATION

On the basis of ecological principles and its good economic efficiency, the pig-grass-fish integration can get better economic efficiency than fish-cum-pig integration. In 1979, Zhangzhuan Aquaculture Brigade performed a productive test. 50 kg of pig exereta (pigsty effluent occupies 50%) or 50 kg of nightsoil can produce 56 kg of terrestrial grasses. 50 kg of net pig manure can produce 112 kg of grasses. The Food Conversion Coefficient of English rye grass and Sudan grass is 18.4–27.5, average 27. Thus, 100 kg of pig excreta through growing grass can be converted into 4.2 kg of herbivorous fish, bring along rearing 1.5 kg of filter-feeding and omnivorous fish, the total yield is 5.7 kg.

According to the tests of Freshwater Fisheries Research Centre, 100 kg of pig excreta can be converted into 2.5 kg of fish, whereas 50 kg of net pig excreta through growing grass can produce 3.2 kg of fish, increasing 133 %. In pig-grass-fish integration, the dominant species are herbivorous fish, which can meet the demand of the market and gain much profit. 100 kg of pig excreta are used to grow grass then to culture fish, resulting in the increment of 2.90 Yuan which is 1.34 times that of fish farming by applying pig excreta directly.

Allocation of pig-grass-fish

In the combination of pig-grass link, the guide line is to facilitate grass planting and manure application. If grasses are planted on pond dykes only, the simple pig sty can be built near the pond; if there is a fodder field beside pond dyke, centralized pig shed near fodder field is taken into consideration. Firstly, the ratio of pig and grass depends on unit-area yield of pasture grasses. Below the critical level of high yield, the more the yield, the more the number of pigs. Secondly, the ratio also depends on the utilization of pond silt. The more pond silt is used, the fewer the number of pigs. Thirdly, it depends upon the quality of soil and pig manure and varieties of pasture grasses. On the basis of production test for ryegrass and sudan grass grown all the year round, the best allocation is 5–6 pigs per mu. The combination of fish-grass link has already been introduced in fish-cum-crop integration.

WEB OF INTEGRATED MANAGEMENT

In order to raise the utilization efficiency of energy, to decrease cost, to increase products and income, several models could be interwoven to an integration web thru longitudinal and transverse nutrients cycles and energy flow in line with the local conditions. There are three kinds of integration web so far.

Multi-level comprehensive utilization web

In the light of different utilization of nutrients by various specialized trader the trades can be connected in series with nutrient cycle and energy flow as a master line. An integration wed can be regarded as a number of connected nutrient cycles and energy flow. The popular form is recycling of animal manures.

That is to say, in recycling of chicken manure, the spilt and undigested chicken feeds can be fully utilized. Sgan Bei Livestock and Aquaculture Farm adopts this kind of integration web. Not only can chicken manure be fully utilized but also it can save about 25kg of fine feeds for pig. increasing net income by 47 yuan/pig. Furthermore, through simple deodorization and sterilization, chicken manure is mixed with fine feeds as composite pellets, which can be used to feed chicken again and then chicken manure can be fed to pig.

In Zhejiang Province, there is a common multi-level comprehensive utilization model, that is, cow-mushroom-earthworm-duck-fish or cow-mushroombiogas-earthworm-duck-fish. Cow dung can be used to grow mushroom; cow urine for fish farming; mushroom soil for earthworm culture; and then, earthworm is fed to ducks, the excreta of which is applied to fish ponds. The levels can be rearranged in line with the local conditions. Thus, different constituents of cow excreta are utilized at different levels or the new constituent produced at the preceding level can be utilized by the next level.

Parallel-cennected web of multi-trades

With fish farming as the center, there may be a parallel connection of multi-trades, such as fish-livestock-crop integration to create sources of feeds and fertilizers. The products and wastes are utilized reciprocally among all trades. Most of integrated fish farms adopt this kind of integration web. E.g. in Xiang Yang Aquaculture Farm, fish farming is combined with pig, cow and geese raising. Animal manures which is self-supporting are used for fish farming and for crops such as pasture grasses, squash, sweet potatoes and soybean and at the same time for cultivation of aquatic plants such as water hyacinth,

Spirodella spp. and Wolffia arrihza which are used as feedstuff for fish, pig and geese. Although soybean is planted on pond dyke in the latter part of a year, water hyacinth can offer 80% of green fodder for fish and pig.

Fish-livestock-crop and agriculture-industry-commerce web

On the basis of parallel connection of fish-livestock-crop integration, all trades develop on both input and output ends. On input end, feedstuff processing industries can be set up and food can be utilized at multilevels; on the output end, the products of all trades needs processing and marketing. The waste of processing factories is used to raise livestock and poultry. This is a combination, which can raise the level of comprehensive utilization of natural resources and the rate of utilization of energy and can increase the products of fish, livestock and poultry resulting in the increment of value, income and job opportunity.

In 1970's, this type of integration web was set up in Dong Hu Fish Farm, Siang Yin County, Hunan Province. In recent years, the management structure and the scale of this web have developed, further, e.g. Helei Fish Farm has developed into an aquaculture, industry and commerce combination. On the axis of abscissae, they integrated fish farming with grass cultivation, pig, cow and duck.

On the axis of ordinates, they process crops, pupae so as to provide green fodder, pupae waste water for fish, livestock and poultry at the end of input; at the end of output, they process the products of livestock. e.g. making preserved egg, slaughtering ducks. The waste of processing factories can be used as feeds for fish, livestock and poultry or fertilizers for fish ponds. The processed products go to market home and abroad. Thus, the economic efficiency can be raised greatly. For example, in 1982, the average net profit was 6.78 Yuan/duck. The people who are engaged in duck raising and egg processing occupy 17% of the personnel of the farm while the net profit accounted for 43%.

APPRAISAL OF ECONOMIC RETURNS OF INTEGRATED FISH FARMING

Appraising methods and procedures

In term of scope, investigation can be divided into general surveying, sampling and typical surveying. In terms of form, live coverage and questionaire through correspondence. Investigation through various methods can get better results, nevertheless, general surveying combining with typical surveying are commonly practised.

Before investigation, the objective, contents and targets ought to be fixed; the precis of investigation be designed; the table of investigation be drawn up and then the surveying can be conducted deeply and widely as planned.

Experiment method

Due to the weak points of investigation, the trial and error procedure should be performed for complement and verification. Experiment on the site should be held before any new decision on technical measurements is adopted. In order to give a correct appraisal, it's necessary to get technical and economic data as complete and accurate as possible.

Data analysis and processing

There are several methods of data analysis:

i. Comparative analysis
ii. Cut and try
iii. marginal analysis
iv. regressional analysis
v. linear analysis

Selecting the best after the appraisal

After the appraisal, the best plan can be selected in line with the local conditions and enterprise capabilities by using the abovementioned methods and overall & systemetical analysis.

Procedure

The procedure of appraisal of economic returns of integrated fish farming technique.

INTEGRATED MANAGEMENT OF FISH AND CROP FARMING

Fish-cum-crop integration is the most ancient and popular pattern of integrated fish farming.

The necessity and the feasibility of fish-cum-crop integration

Why can we combine fish farming with crop cultivation? It's due to the demand of fish feeds and the excessive pond silt. On one hand, abundant silt deteriorates the pond water and yet on the other hand, it is one of high quality manures for agriculture to plant fodder crops which can in turn be used as feeds for fish. Therefore, pond silt is a link between fish and crops in this integration.

Formation and function of pond silt

Large amount of feeds and manures are put into fish ponds year after year. There is a considerable amount of residues and manures which often sink to the bottom. Moreover, the excrements of fish and aquatic animals, the corpse of the aquatic, and alluvial soil incessantly descend to the bottom too. The sedimentary organic material decomposed by bacteria forms a great deal of humus, which in turn combines with sludge on the surface of the pond bottom to form silt.

The thickness of silt varies greatly with many factors. Even in the same pond, the thickness in different locations is different. The mean thickness formed in a high-yielding earthen pond with target yield of 500 kg/mu and with no slope protection is about 10–20 cm, which is about 50–95 m^3/mu/year or 100–190 tons in wet weight.

The proper amount of silt is beneficial to fertilize the pond water whereas the excessive amount is not so good to fish. Microbes propagate rapidly when large amount of feeds and manures are applied. The pH value will decline, the BOD will increase, and gases such as nitrites NH_3, H_2S, CH_4, PH_3, etc. and surplus nitrites will do harm to fish. The median tolerated limit (TLm) value of Silver carp and Bighead fry in 24 hours at the temperature of 25°C is 0.91 mg/l and 0.46 mg/l respectively. Grass carp is more susceptible (Lei Xingzhi Et al 1983). Surplus nitrites will easily induce hemorrhagic septicemia of fish. Silt contains a lot of ichthyopathogen parasites and other harmful creatures. The more the thickness of silt, the more the deterioration of the pond water. It directly affects fish yields and thus, the excessive silt should be removed after the pond is drained.

On the other side of the coin, silt is a manure of high quality which contains several nutritive elements. In Pearl River Delta and Taihu Lake basin, the pond silt is often removed 3 to 6 times every year.

The nitrogen content of pond silt/mu/year is equal to 481 kg of ammonium sulphate. Furthermore, as pond silt is quite complete in nutrition, it contains not only nitrogen but also phosphorus and potassium. Generally, T! of silt is removed per mu per year and that is equal to 1.2 tons of N.P.K. fertilizers. The quick-acting component of silt is equal to 119 kg of fertilizers, which can also serve as additional manure besides the basal one. Silt could increase the thickness of cultivation layer, improve the soil particle structure and strengthen its ability to absorb ions of N.P.K. and the ability to keep the water. It is beneficial to the prevention of fish diseases. Since it is also a slow-acting fertilizer which is beneficial to late crops. In terms of mass's experience, the quick-acting component of 100 kg of silt in dry weight can increase 1 kg of rice. If we only use silt as fertilizer, 100 kg of silt in dry weight can produce 10 kg of rye grass. Provided that 50 m^3 of silt is removed from each mu of fish ponds, all the silt can be used to cultivate 6 mu of rice, and then it could increase about 500 kg of rice; if it is used to cultivate rye grass, the output can reach over 6000 kg/mu.

The yield can be increased by 2.5 times. From investigation data, the yield of per-unit paddy field by using 10–15 tons of silt composted with green grasses is near to the one by using 5 tons of animal manures composted. The cost of digging silt in China is cheaper than the one of purchasing animal manures from outside the fish farm.

The demand of feeds in fish farming

The demand of both commercial feeds and natural food organisms is great in fish farming alone. The feeds and foods sometimes are hard to get and the great amount of supply also affects the balance of the market and ecology.

Besides, the cost of transportation and energy consumption is surprising. The fish pond with the target net yield of 250 kg including 100 kg of herbivorous fish, 100 kg of planktoneater, 50 kg of omnivorous fish needs 1500 kg of aquatic grass, 1000 kg of vegetables, 150 kg of grains apart from manures. In order to meet the demand of feeds and foods in fish farming and to reduce the cost, it's necessary to combine fish farming with crop cultivation.

The feasibility of fish-cum-crop integration

Aquaculture can provide large amount of silt and fertile water for agriculture. There still exists a potentiality of land on fish farms. E.g. the mean pond dyke is three meters wide, the gradient of a slope is 1:1.5–3. The mean area of a fish pond is 10 mu.

The arable area of pond dyke and slope and the area of the water surface is in the ratio of one to five. The arable land might be greater before May with shallow water in the pond.

With extra 0.3 mu of forage field attached to one mu of fish pond plus all the areas available, the average forage field for one mu of fish pond could amount to 0.5 mu or even higher which can provide fodder crops for herbivorous fish and proportionable filter-feeding fish in one-mu fish pond with net target yield of 400 kg. What's more, it's possible to plant aquatic plants on scattered unused surface. In short, it is necessary and feasible to integrate fish farming with crop production so as to fully utilize pond silt, arable land and water surface. As a result of that, the needs of fish feeds could be satisfied wholly or partially.

MAJOR PATTERNS OF FISH-CUM-CROP INTEGRATION

According to crop variety and various cultivation systems, fishcum-crop integration can be divided into four patterns: fish farming integrated with terrestrial crop production; cultivating crops in ponds first, fish farming next; fish farming combined with aquatic plant cultivation and dyke-pond system.

Fish-cum-terrestrial crops integration

All or most part of crops planted in the fodder crop field and corner plots on pond dykes and slopes are used as green fodder for fish and as fertilizers for ponds. This is the most popular pattern in fish-cum-crop integration.

Crop variety

Choose crop variety which are palatable to fish, rich in nutrition, strong in resistance to diseases, easy to manage and has features such as well-developed root, which protects the slope, etc. If it serves as straw manure, it should be easy to decompose. Besides, the average yield of some leguminous plants such as Trifolium repens, T.pratense, Medicago sativa, Astragaluo sinicus can reach 5000–7500 kg/mu. These grasses serve as both feeds and fertilizers. The average yield of some gramineous plants such as Pennisetum purpureum, Phalarosarundinacea, Pennisetum alopecuroidescross, P. purpureum can reach over 10,000 kg/mu. The seeds and young crops of other grains such as barley, wheat, maize and rice are also palatable foods of fish. The tender and juicy vines and leaves of sweet potatoes Ipomoea batatas and squash Cucurbita are pulverized as fish feeds for Grass carp in Wuxi area. The tuber and squash, etc. are cooked to feed feed-eaters.

COLLOCATION OF FISH FARMING AND CROP CULTIVATION

The ingestive variation of fish is caused by the growth of fish and environmental conditions. Among the environmental conditions, the seasonal change of water temperature is the main factor. The production period, therefore, has to be collocated with this variation of fish ingestion so as to synchronize the daily production of fodder grass with the daily ingestion of fish. The period of the lowest ingestion amount of cultivated fish in Changjiang Drainage is between Feb. and Mar. and between Nov. and Dec. in a year. The peak is between Jun. and Sept. accounting for 50% of the annual total amount. The production mode of English Ryegrass and Sudan grass nowadays coincides with the demand of fish feeds.

Ryegrass is sowed in Sept. every year and transplanted in Oct. and mowed in Dec. in Changjiang Drainage. It reaches the peak production between Apr.

and May. The integrated fish farm can supply itself sufficiently by the production of Ryegrass. Sudan grass is sowed in mid Apr. and mowed for the first time when it grows to 50 cm high just before the withering of rye grass. It reaches the peak production between June and Sept. The daily output is about 100 kg/mu. Based on the experiments the annual production of 2 grasses can reach about 15,000 kg/mu if it's well managed. The other ways of collocation are as follows:

A. To intercrop gramineous grasses: Ryegrass, Sudan grass and leguminous fodder grasses.
   The unit area output of two families of fodder grass can be increased and the quality can be improved.
B. To adopt ensiling method if there is a surplus of ryegrass or to pelletize ryegrass with other materials, so as to be supplied during crop change in June as supplements.
C. To adopt sowing and transplanting by stages and to mow the grass in turn.

In Changjiang Drainage the seeds of Sudan grass are sowed by stages in April ahead of the usual time and can be mowed in mid May. In order to maintain the fertility of the land, pond silt must be fully utilized with some additional organic fertilizers apart from planting leguminous grass, and at last, the grasses should be harvested with stubble remaining in the field.

Stocking models of fish in fish-cum-crop integration

The main or only source of feeds and fertilizers is the pasture grasses cultivated on the farm. The spp. stocked should be grass carp and Wuchang fish megalobrama spp. as major spp. which occupy 60–70%. Silver carp & Bighead as minor spp. which occupy about 20–30% assorted with a few of omnivorous fish which occupy about 10%. The omnivorous fish can not only utilize residues and detritus but also can clean the fish pond for herbivorous fish but the ratio should be limited within 10%.

In order to get natural food organism ready for Silver carp and Bighead after stocking, it is necessary to mow the grass and make a compost to fertilize the pond water or stock Silver carp and Bighead half a month later after stocking of Grass carp. If the farm produces more compost, the proportion of omnivorous fish may be increased a little.

The use of pond silt in fish-cum-crop integration There are 2 methods to use pond silt in accordance with crop strains and cropping system:

a. Pond silt is directly used as base manure for Ryegrass, etc. The silt should be harrowed and smoothed after it dries a little and then the seeds can be sowed. If it's a transplantation, there's no need to harrow. The seedlings could be interpolated, economizing on much labour and time. If the silt is used as additional manure, it should be topdressed to the roots of the plant. In summer, silt will be bailed with water to the fields. This method is beneficial for the silt to release or diffuse nutrient elements and detritus. It also benefits increasing of dissolved oxygen in the bottom layer of water.

b. Making compost with pond silt and grass

Usually, there's a surplus of pond silt in winter whereas there's a shortage of pond silt in summer. Therefore dig out a pit, 5 meters square with a depth of 1 m near the crop field and then put in pond silt and grass (straw manure and stable matted grass are better) to make compost and cover the pit with mud. After fermentation, the effectiveness of manure is increased. The compost is usually used as base manure, sometimes mixed with water as topdressing.

It's safer to use pond silt as manure. The applying amount is not necessarily limited, with an average rate of 5–15 tons/mu or even a thickness of 35 cm on the soil.

ECONOMIC EFFICIENCY OF FISH-CUM-CROP INTEGRATION

To increase sources of feeds and fertilizers

This pattern of integration could provide a considerable amount of feeds for fish farming. The source of feeds is stable. They are much cheaper and of high quality. The cost could be reduced by one third.

To save energy

The energy consumption of transportation of purchasing feeds and fertilizers could be reduced.

To rationally utilize the labour

Fish farming is a seasonal work. The input of labour varies greatly. The labour in slack season can be used for crop production. Part time worker can be fully utilized. This offers more jobs and increases the income

To set up reasonable ecological system

Pond silt is used to plant fodder crops which in turn are used to feed fish. Large amount of fish could be produced and pond silt is accumulated again. It's a cycle which can fully utilize the sunlight, land, pond silt and fertile pond water and can improve the ecological condition of a pond. According to some data, grass planting on pond dyke can reduce the erosion of soils by 57%.

Fish-cum-crop integration in line with the local conditions means to combine the Chinese traditional aquaculture with modern technology.

"Rotation" of fish and grass

Owing to the non-continuity of fish farming production, most of nursery ponds and grow-out ponds have been left fallow for a certain period of time, e.g. a yearling rearing pond has half a year of fallow from Nov. to June next year. During this period, fish ponds can be used to plant green fodder or green manure crops and then to culture fish. This could lead to the full utilization of fish ponds.

The method of "rotation" of fish and grass

The "rotation" of fish and grass can be divided into single pond rotation and multi-pond rotation.

Single pond rotation

That means the "rotation" of fish and grass in the same pond. This method is usually adopted after transferring fingerlings to other ponds.

Drain the pond, expose its bottom to the sun, trim the pond dyke, smooth the bottom, dig a ditch for drainage and then sow seeds of rye grass or barnyard grass or rice, etc. on the bottom and the slope. The seeds of rye grass are broadcasted in Nov. at a rate of about 2 kg/mu while the seeds of barnyard and rice are generally sowed in late Apr. or early May in Jiangsu Province. In order to extend the growing period germinated seeds could be sowed in the 1st 10 days of Apr. at a rate of 5–6 kg/mu. After sowing, get rid of birds and control the water depth and adopt different field management towards different plants. Rye grass is a dry crop which can't endure waterlogging. Barnyard grass seedlings like to grow in shallow water with a depth of about 5–6 cm. During their growth period, they can be mowed as green fodder crops. Ryegrass can be mowed four times, with the yield of 3000–5000 kg/mu; barnyard grass and rice can be mowed once with the yield of about 2000 kg/mu. The last ratooning plant of the green fodder or green manure crops should be submerged for fermentation as manure. But the amount of grass per square meter should be controlled within 4–5 kg. Then add water to about 2 meters high. In this way, it won't cause lack of oxygen in the pond water. 11–15 days later summerlings are stocked. After stocking, keep close watch over the water quality and fill the pond with fresh water or change the pond water in time accordingly.

Barnyard and rice can be mowed and submerged by stages on the basis of their growth and then different summerlings can be stocked by groups.

Multi-pond rotation

It needs a set of fish ponds, of which, a few are used to grow grass or green manure crops. Here are two ways. Because of short period of fallow of grow-out ponds, a few of fish ponds are drained first and planted with grass as early as possible but stocked with fish later than the other ponds, thus, extending the growing period of the grass; the rest are drained later but stocked with fish first. The fish ponds are rotated to do this each year. If there is a food constraint for fish, one or two fish ponds could be used to grow crops to provide feedstuff for fish in the other fish ponds. The fish ponds are rotated to grow grass each year.

THE EFFECTS OF "ROTATION" OF FISH AND GRASS

The system of rotation of fish and grass can fully utilize the productivity of fish ponds to provide both green fodder crops for fish and green manure crops for fish ponds to propagate large amount of natural food organisms.

According to the practices of Suzhou Municipal Fish Farm, the phytoplankton can reach 3×10^8 ind/1 at the peak time. The growth of summerlings can reach 1.8–2.5 mm/day in the initial month. If the yield of barnyard grass in a pond is 5000kg, 40kg of grass can be transformed into 1 kg of fingerlings averagely. It can save 0.5–0.8 kg of marketable feeds and reduce 10% of the cost.

In this system, crops utilize the nutrients of pond silt, whereas crop nutrients, which return back to the pond after submergence, not only improve the water quality but also increase the soil fertility. Practices prove that the fish yield increases after submergence of crops, e.g. Suzhou Municipal Fish Farm has adopted this method to nurture fingerlings since 1979. In 1982, the average yield increased from 150 kg/mu to 391 kg/mu, increasing by 1.6 times.

Fish farming integrated with aquatic plant culture

In a network of rivers like lower reaches of Changjiang River, a fish farm is often near lakes, rivers or water-logging area or in the vicinity of inlet and outlet of irrigation canal. These water bodies are rich in nutrients, esp. effluents from cities and fish farms. In order to utilize these water resources, fish farmers in the southern part of China often culture aquatic plants in these water bodies. The principal aquatic plants are water hyacinth Eichhornia crassipes, water lettuce Pistia stratistes and water peanut or alligator weed Alternanthera philoxeroides, so-called three Ap's; the secondary are duckweed such as (Spirodela polyrhiza) and Wolffia arrhiza and Lemuna minor.

The ways of utilization of aquatic macrophytes

The output of aquatic macrophytes is the highest among the green fodder crops for fish. The yield could be about 15 ton/mu, even above 25 ton/mu. The aquatic macrophytes grow too fast and it causes a lot of trouble in the Tropics and sub-Tropics. However, the nutrients of aquatic macrophytes are higher.

Water hyacinth is called "King of Aquatic Plants". So far as protein produced from per unit area is concerned, it is 6–10 times that of soy bean. Aquatic macrophytes are easy to manage with less labour and lower cost. It's said that one labour can manage 50 mu of 3 aquatic plants and can produce 13.1 ton of crude protein in half a year. The cost of one ton of three aquatic plants (including wages) is only about 1 yuan.

THE METHODS OF UTILIZATION

Three Ap's directly serve as fish feeds

Three Ap's are processed in different ways to turn into different sized feeds for a variety of fish spp. in different size.

To nurture fry, 3 Ap's should be mashed into grass paste, and the residues of leaves must be filtered out; then the paste can be sprinkled to the whole nursery pond. Table salt, 2—5% of the weight of the plant should be added to the water hyacinth paste to decrease the toxicity of saponin. The equivalent ratio of three Ap's and soy bean in fish farming is 17.5—25 : 1. It needs 5 kg of soy bean to nurture 10,000 ind of summerlings. We can use 87.5—125 kg of 3 Ap's instead. The cost of soy bean is 13–18 times that of three Ap's. Fry will grow fast with higher survivability if three Ap's are used as fish feeds. For example, Zhuang Aquaculture Brigade, Wuxian Country stocked 80,000 fry in 1977. After 16 days of nurturing, the transferred size was 3,33 cm, with 94.5% survivability from the pond fed with water hyacinth paste while the transferred size was 2.93 cm with 92.8% survivability from the pond fed with soybean milk.

To rear fingerlings three Ap's are esp. good for Silver carp and Bighead. 3 Ap's should be mashed into grass paste too but it's not necessary to remove the residues.

To rear adult fish, with herbivorous fish of different spp. in different sizes as major spp. three Ap's are often pulverized by green fodder crops pulverizer and then fed to the fish.

According to the experiments, about 45 kg of three Ap's could be converted into 1 kg of fish. The macrophytes per-mu can be converted into more than 400 kg of fish. If rice and wheat brans are used as feeds, it will need 1600 kg of brans to produce the same amount of fish. The cost of three Ap's is about 10% of that of brans, but three Ap's are less effective if they are not process. Moreover, since the contents of N.P. and K in three Ap's are , the grass paste could serve as manure in fish ponds also.

Three Ap's" - livestock, poultry-fish

"Three Ap's" are also palatable food for various animals in integrated fish farms. They do not need processing or just a simple processing, so the rate of utilization is very high. 900—1000 kg of three Ap's can rear one piglet to an adult with a body weight of 60–70 kg with a little amount of wheat and rice brans. The excreta of one pig can be converted into more than 40 kg of fish. In Helei Fish Farm, water hyacinth is fed to ducks at a rate of 150 g/duck/day with a little amount of wheat and rice brans. This can save 10,000 Yuan in feeding 22,000 ducks in half a year. The average excrement of each duck is about 52 kg and can be converted into 3 kg of fish. Thus, three Ap's become fish yield through two trophic levels (i.e. feeding levels).

DYKE-OOND SYSTEM

This is a special pattern of fish-cum-crop integration. On some farms, much of the broad dykes is devoted to economic crops:mulberry bushes, sugarcane, fruit trees, tea and rape, etc, or crops, vegetables and grasses. With different crops, this system can be called "mulberry plotfish pond"; "sugarcane dyke-fish pond"; "fruit tree dyke-fish pond" "rice paddy-fish pond", etc. Among these, mulberry plot-fish pond is more popular. Most of dyke-pond systems are distributed in Pearl River Delta and Taihu Lake basin. According to the historical records, mulberry cultivation and fish farming could be traced back to the 5th century B.C., but they did not form a well-linked dyke-pond system until the 16th century. The dyke-pond system in Taihu Lake basin is similar to that in Pearl River Delta, which is related to the common factors of the nature and the society. Take mulberry plot-fish pond system as an example:

Ecological structure of "mulberry plot-fish pond" system This system has a perfect ecological structure. It includes mulberry cultivation, sericulture, silk extraction, and fish farming with silkworm faece, pupae and waste water. Mulberry is the producer; silkworm is the first consumer; fish are the second ones land ecological system. In fish ponds, there are four ways of energy flow:

i. Silkworm faeces are directly taken in by fish and part of detritus can be filtered by filter-feeding fish.

ii. The inorganic nutrients in silkworm faces are utilized by phytoplankton & heterotrophic bacteria and the biomass of phytoplankton and bacteria in turn are eaten by filter-feeding fish directly or indirectly.

iii. The leftover of input and fish faces are decomposed by hydro microbes, releasing inorganic nutrients; and then, the same process occurs as in (ii).

iv. At the same time, pond silt which is composed of all kinds of sediments returns to the pond dyke and the new material cycle begins.

THE MATERIAL LINK OF "MULBERRY PLOT-FISH POND"

The link of energy flow in "mulberry plot-fish pond" system is pond silt and feeds.

Pond silt

Pond silt is the main source of manure for crops on the pond dyke. Each mu of fish ponds can provide pond silt for 1–2 mu of plants on dyke, usually for direct application but making compost in paddy-fish ponds in Taihu Lake basin. In winter pond mud (much mud, less water) is removed from fish ponds after draining and applied between mulberry bush lines on pond dyke or bailed (with a little water) to pond dykes and spread evenly after dry. After that, winter crops are intercropped between mulberry bush Lines.

In summer and autumn, liquid silt known as "nihua" (less mud, much water) is bailed to the pond dyke after mulberry leaves are picked 1–2 times, which is practised 2–3 times in Pearl River Delta and 5–6 times in Taihu Lake basin respectively every year to increase the soil fertility and when the base thickness of the soil of mulberry plants increase by 5–6 cm, it is beneficial to next crops. The mud on pond dykes are washed into fish ponds by rains; pond silt is supplied to pond dykes with nutrients. The process runs again and again.

Feeds and fertilizers

Mulberry plots can supply feeds for fish directly or indirectly:

a. To provide feeds directly, including pasture grass, vegetables and mulberry leaves.

   On mulberry dykes, grass and vegetables can be planted after autumn leaves being picked; grasses and vegetables could be harvested at a rate of 3000 kg/mu and they can be converted into about 100 kg of herbivorous fish and 40 kg of other fish.

b. To provide feeds indirectly.

The first one is silkworm wastes which are a mixture of silkworm faeces, worm sloughs and mulberry leave residues. Silkworm wastes are rich in nutrients: organic material 87%; N 2.2–3.5%; P_2O_5 2.0–2.5%; K_20 1.5–2% and several trace elements but with less moisture, so the nutritive contents of silkworm wastes is higher than that in any kind of livestock or poultry manures. They serve as both fertilizers and feeds. One mu of mulberry plot can produce 2400–2500 kg of mulberry leaves, even 5000 kg. 100 kg of mulberry leaves when fed to silkworm can produce about 30–50 kg of silkworm wastes; therefore, 1250 kg of silkworm wastes can be obtained from sericulture with the mulberry leaves from one mu of mulberry plots. The conventional FCC of silkworm wastes in fish ponds of polyculture is 8. Hence, 1250 kg of silkworm wastes can be converted into over 150 kg of fish. The next one is pupae, which are by-products of silk extraction of cocoons. The protein content of pupae holds 55.8%, crude fat 29.1%. Pupae therefore are good feedstuff for feed-eaters.

Its FCC is 1.5–2. In Pearl River Delta, the production of cocoons from one mu of mulberry field is between 160 and 175 kg out of which about 130 kg of pupae can be obtained, and used as fish feeds, producing 90 kg of feed-eater spp. and their assorted spp. The third one is waste water with large amount of protein detritus and soluable protein, which is obtained when the cocoons are steamed and processed at the factory. According to practices of the masses, 200 kg of waste water can be transformed into 1 kg of fish. More than 2500 kg of waste water can be obtained from the processing of cocoons which are gained from sericulture with mulberry leaves of one mu. Thus, 15 kg of fish can be produced. When the ratio of dyke and pond is 1:1, the total products (except pupae) from dyke can meet the demand of feedstuff for the fish pond with net fish yield of 300 kg.

INTEGRATED MANAGEMENT OF FISH-LIVESTOCK-POULTRY FARMING

Fish-Cum-Duck Integration

The model of fish-cum-duck integration is rather common in China. Both land and water surface are the habitats of ducks. Fish-cum-duck integration is to utilize the biological relationship between fish and duck, which is of mutual benefit. It's not only beneficial to the fattening of ducks but also beneficial to the fish farming by providing more organic manures to fish. Thus, the fish yields can be increased. It's apparent that fish-cum-duck integration could result in good economic efficiency of fish farms.

The history and status quo of fish-cum-duck integration

Duck raising in fish ponds is an ancient practice in Asia and Europe and this has long been practised in China too. The ancient farmers probably didn't know the advantageous effects of duck raising on fish culture. It's more likely that this integrated production pattern was formed incidentally on the basis of swimming propensity of ducks.

In 1934, the German scientist Probst conducted scientific experiments of integrated fish-cum-duck farming for the first time. He raised ducks in Common carp fish ponds and found that 0.9–1.7 kg of fish could be increased by raising one duck. Since the original output of fish in the experimental fish pond was rather low at that time, the increment of fish yields was conspicuous after ducks were raised in the pond; however, the outbreak of World War II kept the results unemployed. After World War II, the food was in shortage, especially animal protein. The short supply of animal protein urged the development of the intensive commercial fish farming. Hungary, Czechoslovakia, GDR began to conduct integrated management experiments of fish-cum-duck in large scale early or late from 1952 to 1955. So far, integrated management pattern of fish-cum-duck has chiefly been practised in China, Hungary, GDR, Poland and Soviet Union. Fish-cum-duck integration has developed into a fixed model of integrated fish farming. In recent years, this model either on the scale aspect or on the managerial aspect has been developing very rapidly, especially in the area of a network of rivers like Jiangsu and Zhejing Provinces.

Beneficial interactions in fish-cum-duck polyculture

From biological standpoint, a fish pond is a semi-closed biological unit. In fish ponds, there are many aquatic animals and plants, most of which could be used as natural food organisms of fish; some are predators of fry and fingerlings or detrimental to fish but can be utilized by ducks. If duck raising is conducted in fish ponds, the water surface can also be put in full utilization. Fish ponds provide ducks with an excellent environment, which prevents them from infection of any parasitic and other diseases. Ducks like to eat juvenile frog, tadpole, larvae of dragonfly.

That means to eradicate predators for fry and fingerlings. Furthermore, the protein content of these natural food organisms of ducks is high. Therefore, duck raising in fish ponds can appropriately reduce the demand for protein in duck feeds. For ducks raised in pens, the digestable protein content in duck feeds must be 16–18%, even higher up to 20% while for ducks raised in fish ponds, the digestable protein content might be reduced to 13–14%. This can save 200–300g available protein each duck, an equivalent to 2–3% of duck feeds.

If the water surface is employed to raise ducks, the droppings of ducks directly go into water bodies providing C, N and P elements frequently. Thus, the biomass of natural food organisms in fish ponds can be increased. Here are two merits:

First, the droppings of ducks go directly into the pond water so that the availability of manure will not be lost at all; second, the "manuring" conducted by ducks on the surface is more homogeneous without any heaping of duck droppings. For this reason, duck raising in fish ponds can promote the growth of fish and increase the fish yields; in the meantime, it can naturally eliminate a series of problems which might be caused by pollution of duck excreta in duck pen.

Duck droppings are good organic manures in pond fish culture. The organic substance content in duck droppings are 3–5 times that of human excreta, inorganic substance such as N, 1.5–2.2 times; K_2O, 2.6–3.1 times. It shows that the quality of duck excreta is higher than that of human excreta. However, the quality and quantity of duck excreta are determined by duck species, feeds given, culturing management as well as climatical conditions, etc. In Europe, the stocking rate of ducks is generally 300–500 ducks/ha in summer and each duck produces about 7 kg of droppings during the fattening period of 36 days. If 500 ducks are raised, the total duck excreta would reach 3000–3500 kg in that period. The moisture content in duck excreta holds 56.6%; organic substance, 26.2%; C,10%; P_2O_5, 1.4%;N,1%; K_2O, 0.62%; Ca, 1.8% and some other elements. The annual excreta of each egg-laying duck is 7.5–10 kg in dry weight, about 70 kg in wet weight. Each egg-laying duck raised in Helei fish farm, Wuxi, produces 40–45 kg of manure in wet weight per annum.

Besides the above mentioned, fish-cum-duck integration is beneficial to fish farming:

1. It can fully utilize the uneaten feeds because 10–20% of feeds are spilt by ducks to the ground. Based on the investigation, the fine feeds lost is 23–30g per duck per day or even higher. These spilt feeds and leftover falling directly into the pond water or flushed into fish ponds are utilized as fine feeds for fish resulting in achieving relevant fish yields. Furthermore, it reduces the wasting of duck feeds.
2. It can promote the recycling of nutrients in pond ecological system. The water surface is usually the duck's habitat. In shallow places, a duck often puts its head into the pond bottom and turns the silt to search for benthos. By virtue of this digging action, the nutritional elements deposited in the pond humus will diffuse. It's beneficial to the material recycling in the pond.

3. Ducks are the volunteer aerators. The swimming, playing and chasing of ducks in a pond more or less play a role in aeration, disturbing the smooth surface of the pond in reality.

Barash et al (1982) carried out an experiment on fish-cum-duck integration, rearing Common carp, Tilapia, Silver carp and Grass carp in an experimental pond of 400 m^2. Ducks were fed with nutrient-balanced feed pellet. At the beginning of the test, duck manure was applied at a rate of about 85 kg/ha in dry weight.

At the end of the test, at a rate of 95 kg/ha in conformity with the growth of ducks. There was no other input of manures and feeds except duck excreta and spilt feeds. Compared with the duck raising in pens, the growth rate, feed efficiency and vitality and cleanness of eiderdowns and skin of the ducks raised in fish ponds are better than that in duck pens. Daily fish yield reached 38.5 kg/ha.

Duck raising in fish ponds compared with duck raising in duck pens has three advantages:

1. The feed efficiency and the body weight of each duck increased. The higher feed efficiency also implies the spilt feeds uneaten by duck could be utilized by fish. Thus, the food conversion rate of fish-cum-duck integration was reduced from 3.84 to 2.64.
2. The survival rate of duck raising in fish ponds increased by 3.3% because fish ponds provided a clean environmental conditions for ducks.
3. Without any other feeds, only the droppings and leftover of ducks made the daily fish output of 36.5 kg/ha. However, someone thinks that if fish and ducks are raised together in the same pond, ducks will eat small fish. Based on many years practices in Helei fish farm and the introduction of duck raising experts, it's easy for ducks to swallow a fish with a body weight below 4 g. Fish with a body weight above 5g have a strong ability to swim and can get away from ducks. In Helei Fish Farm, the recovery rate does not decline when Silver carp summerlings are stocked in fish-cum-duck ponds. Of course, the fry ponds and yearling ponds are not suitable to raise ducks because fish are too small and in high stocking density. Duck raising is generally practised in fingerling ponds or grow-out ponds.

THE RATIO OF FISH-CUM-DUCK INTEGRATION

In fish-cum-duck integration, the result of fish farming has much to do with how to raise ducks. There are three types of farming practices:

Raising large groups of ducks in open water

This is "grazing" type of duck raising. Average number of a group in grazing is about 1000 ducks by one worker. The ducks are generally let loose to go grazing in rivers, lakes and reservoirs during the day, utilizing natural food organisms, but are kept in pens at night. This method is advantageous to large water bodies for promoting fish proliferation and can be considered as an integrated management model for large water bodies, such as lakes and reservoirs, etc. As for those fish pond farms, this method of integration can not effectively utilize the duck manures except the increment of economic benefit from duck raising.

Raising ducks on pond shore

A relatively large duck shed including a workshop for the administration is constructed in the vicinity of fish ponds with cemented area of dry and wet runs outside. The average stocking rate is about 4 ducks/m^2. The dry and wet runs are being cleaned once a day. During cleaning, the sluice of the wet run is opened to allow organic manure to be flushed into fish ponds through a manure ditch. After this, the sluice is closed and the wet run is filled with fresh water. This method has the advantages for centralized management and adopting mechanisms, but is unable to fully utilize leftover and undigested feedstuff in duck excreta. The effects of leftover and the direct value of feeds more or less are lost. It's also unable to take advantage of the symbiosis of duck and fish.

Raising ducks in fish ponds

This is the common method of integrated fish-cum-duck farming. The dikes of grow-out or 2-year-old fingerling ponds are partly fenced to form a dry run and part of the water area or a corner of the pond is fenced with used material to form a wet run. The net pen is installed by 40–50 cm above and below the water surface respectively in order to save the net material. In this way, fish can enter the wet run for food while ducks can not escape under the net. In a large pond, a small "island" is constructed at the center of the pond for installation of demand feeding facilities. The stocking density currently practised in China is higher than that practised in other countries, averaging 4.5 individual/m^2 of pen shed including the dry run and 3—4 ind/m^2 for the wet run.

In the early phase of integrated management of fish-cum-duck, ducks went everywhere in fish ponds to get food; now this pattern has already been improved. The duck raising area has been set up to connect the duck shed, the dry run and the wet run. Whether fish-cum-duck integration succeeds or not primarily depends on technical measures of duck raising. Both meat ducks and egg laying ducks can be raised in fish ponds. In summer, the 14-day ducklings are accustomed to the life on the water surface. Meat ducks grow fast. The good stock can reach to the marketable size of 2 kg in fish ponds within 48—52 days; the stocks which grow slowly can reach to the marketable size within 55—56 days. Ducks should be on the market as soon as they reach the marketable size or they will lose their feathers, resulting in decrease of food efficiency, body weight and commodity value.

The number of ducks to be raised in this type of polyculture depends on the quantity of duck excreta which in turn are determined by the duck species, quality of feeds given as well as the method of raising them. In raising Beijing ducks, about 7 kg of duck excreta per duck can be obtained during 36-days fattening period. The egg laying Shaoxin ducks raised in Wuxi produce 42.5—47.5 kg of manure per duck per annum, while the hybrids of Shaoxin and Khaki-Combell ducks produce more than 50 kg per duck. The stocking rate of ducks also depends on the climatical conditions and the stocking ratio and density of the various fish species polycultured in the pond. In Europe, the stocking rate is generally 500ind/ha. As a result, the increment of fish yield will be 90 kg/ha. In tropic and sub-tropic zone, Woynarovich (1980) recommended that the stocking rate should be 2250 ind/ha. In Hongkong, the optimum stocking rate is 2505—3450 ind/ha. In Wuxi, 2000 ind/ha. If the meat ducks are raised in fish ponds, the number should be reduced in view of larger quantity of excreta. In Taihu district, polyculture of 7—8 species of fish is practised in fish ponds, in which the stocking ratio of various species of fish remains unchanged when ducks are raised. If the number of ducks exceeds 3000 ind/ha, then the filter-feeding fish and omnivorous fish could be increased and herbivorous fish could be reduced.

The stacking of organic materials won't take place in this type of integration so long as the stocking rate of ducks is proper and the excreta amount does not exceed the transforming power of a fish pond. Ducks swim loosely in the wet run to search for food, their excreta drop evenly into the wet run and the effects of fertilization of duck droppings expand to the whole pond through winds and waves.

Economic efficiency

The integrated management of fish-cum-duck farming especially raising ducks on the surface of fish ponds is an economically efficient farming practice. In 1980, Helei Fish Farm conducted a comparative test between pond no. 13 and pond no. 21: Two ponds were adjacent with the same size about one ha.

There were pigsties set up on the dyke between the two ponds. Pond no. 13 was stocked with 2207.5 kg of fingerlings and 1900 egg laying ducks without any input of manure; Pond no. 21 was stocked with 2188.5 kg of fingerlings only. The species and size of fish and the feeds given in Pond No. 21 were the same as those in Pond No. 13 but with input of manure at a rate of 20,000 kg/ha/year. The output of Pond No. 13 was 12,234 kg while Pond No. 21 was 10,464 kg.

From the above mentioned comparative test, if the stocking rate of egg laying ducks is 1830—1920 ducks/ha, apart from economizing input of 300000 kg of manure, per ha fish yields can be increased by about 17% over ponds without duck integration. Based on the data and calculation, the average rearing period of ducks is 10 months in a year and about 2.5-5.5 kg of fish might be produced by raising one duck. According to the test in Helei Fish Farm, about 2.5 kg of fish could be produced besides about 200 eggs, (260—300 eggs for the hybrids) by raising one duck in fish pond. Helei Fish Farm raise 20000 ducks every year, providing 850000 kg of duck manure. In 1980, the farm got a net profit of 42000 yuan from duck raising. There were 48 farmers in the farm. The average income was 881.25 yuan per capita.

It's impossible to get an accurate economic analysis on fish-cum-duck integration because the investment, production cost and the yields of duck and fish varies in different countries; even in the same district, fish and duck species, stocking densities, quality and efficiency of feeds, rearing management and climatical conditions differ in thousand ways.

It can be explained by an example:

In 1981, Helei Fish Farm raised 22000 ducks. Apart from providing 1,000,000 kg of duck manure and large amount of leftover to fish ponds, it harvested 212,695.9 kg of duck eggs, 6059 kg of duck meat, thus providing the market with 24315 kg of available animal protein which is equivalent to 215560 kg of Grass carp protein. The total annual income was 57740 yuan including 10800 yuan from ducks, 42000 yuan from duck eggs, 4940 yuan from duck excreta for fish farming and it holds 24.62% of total profit of the farm.

From the view point of input-output relationship, integrated fish farming with ducks is considered to be the best model in integration of fish, livestocks and poultry. The economic efficiency of fish-cum-pig integration from micro-economic stand point is generally not so high and its profit is low. In the case of integrated fish farming with poultry, fish-cum-chicken integration lacks symbiotic relationship, which exists in the case of integrated fish-cum-geese farming, but the egg-laying rate of geese and the market demand are rather low. By comparing the protein input and output in integrated fish-cum-duck farming, it's found that the production of 1g of egg protein from Shaoxin ducks requires 5.53 g of feed protein whilst 1g of dairy protein from cows needs 5.55g of feed protein. They are similar in weight; however, it is relatively easier to raise ducks than cow and the economic efficiency and income of fish-cum-duck far exceeds that of fish-cum-cow, e.g. in 1981, a worker in Helei Fish Farm produced 292.6 kg of protein from cows while his counterpart produced 506.6 kg of protein from ducks which is 73% higher. The net profit per capita in fish-cum-cow farming is 1067.9 yuan and 1427.08 yuan in fish-cum-duck farming, which is 33.6% higher.

Integrated management of fish-cum-duck farming can be further developed to achieve higher economic efficiency by utilizing the natural water body to cultivate high-yield aquatic plants as vegetable feeds of ducks and by utilizing the wastes of integrated fish farming and from city proper to grow earthworm as animal-based feeds for ducks. At the output end, the products such as fish, ducks and eggs could be further processed before marketing, thus, the economic efficiency and income can be considerably raised either from the angle of energy and nutrient source or economy.

7

Biofertilizers in Farming

APPLICATION OF BIOFERTILIZERS

- Seed treatment or seed inoculation
- Seedling root dip
- Main field application.

Seed Treatment

One packet of the inoculant is mixed with 200 ml of rice kanji to make a slurry. The seeds required for an acre are mixed in the slurry so as to have a uniform coating of the inoculant over the seeds and then shade dried for 30 minutes.

The shade dried seeds should be sown within 24 hours. One packet of the inoculant (200 g) is sufficient to treat 10 kg of seeds.

Seedling Root Dip

This method is used for transplanted crops. Two packets of the inoculant is mixed in 40 litres of water. The root portion of the seedlings required for an acre is dipped in the mixture for 5 to 10 minutes and then transplanted.

Main Field Application

Four packets of the inoculant is mixed with 20 kgs of dried and powdered farm yard manure and then broadcasted in one acre of main field just before transplanting.

Azospirillum/Azotobacter

In the transplanted crops, *Azospirillum* is inoculated through seed, seedling root dip and soil application methods. For direct sown crops, *Azospirillum* is applied through seed treatment and soil application.

Phosphobacteria

Inoculated through seed, seedling root dip and soil application methods as in the case of *Azospirillum*. Combined application of bacterial biofertilizers.

Phosphobacteria can be mixed with *Azospirillum* and *Rhizobium*. The inoculants should be mixed in equal quantities and applied as mentioned above.

Rhizobium

For all legumes *Rhizobium* is applied as seed inoculant.

Points to Remember

- Bacterial inoculants should not be mixed with insecticide, fungicide, herbicide and fertilizers.
- Seed treatment with bacterial inoculant is to be done at last when seeds are treated with fungicides.

Table. Biofertilizers Recommendation (One Packet - 200 g)

S. No.	Crop	Seed	Nursery	Seedling Dip	Main Field	Total Requirement of Packets Per ha
1.	Rice	5	10	5	10	30
2.	Sorghum	3	-	-	10	13
3.	Pearl millet	3	-	-	10	13
4.	Ragi	3	-	5	10	18
5.	Maize	3	-	-	10	13
6.	Cotton	3	-	-	10	13
7.	Sunflower	3	-	-	10	13
8.	Castor	3	-	-	10	13
9.	Sugarcane	10	-	-	36 (3 split)	46
10.	Turmeric	-	-	-	24 (2 split)	24
11.	Tobacco	1	3	-	10 g/pit	14
12.	Papaya	2	-	-	10	-
13.	Mandarin Orange	2	-	-	10 g/pit	-
14.	Tomato	1	-	-	10	14
15.	Banana	-	-	5	10 g/pit	-

Phosphobacteria: The recommended dosage of *Azospirillum* is adopted for phosphobacteria inoculation; for combined inoculation, both biofertilizers as per recomm-endations are to be mixed uniformly before using.

NATURAL AND LIVING ORGANIC IN FERTILIZERS

Biofertilizers are the new thing and yet they have been used for centuries. The fertilizers are natural and living organic substances that carry beneficial fungi. These fungi colonize the roots or soil around plants and help increase the ability to utilize and uptake nutrients. The effect can come from using compost or other organic material that has mychorrhizal fungi growing in the compound. Other beneficial bacteria are rhizobium and ectomychorrhiza.

Table. Rhizobium (Only Seed Application is Recommended)

S. No.	Crop	Total Requirement of Packets Per ha
1.	Soybean	5
2.	Groundnut	5
3.	Bengalgram	5
4.	Blackgram	3
5.	Greengram	3
6.	Redgram	3
7.	Cowpea	3

Olive Tree Nutrient Needs

Olive trees do not need a high amount of nutrients to perform their functions and produce fruit. The more important requirement is water. Olive trees grow in nutrient poor, rocky or gravel-laden soils. The plants produce excess vegetative growth when they are given extra nutrients. These long shoots become unproductive even when pruned to force branching.

Organic fertilizers or soil amendments like legume cover crops, decompose slowly and deliver a slow release of nitrogen to olive trees that is only 6 to 10 per cent and will not cause a burst of growth.

New Olive Trees

Young olive tree seedlings benefit from biofertilizers. A study was cited in the In Vitro Cellular and Developmental Biology publication in 2007. The study titled, "Microp-ropagation of Olive (Olea europa L) and Application of Mycorrhiza to Improve Plantlet Establishment," focused on three French olive varieties where mychorrhizal fungi were introduced to microplantlets grown in mircropropagation chambers.The naturally occurring fungi create a mutually beneficial symbiotic relationship with the roots of the olive plants. They increase nutrient uptake, increase the root size and protect the plant against biological threats. This means the fungi act as biofertilizers and also bioregulators. The plants had a higher long-term survival rate and acclimated to transplant better than those plants not inoculated.

Biofertilizers and Phosphorus

One of the most common nutrient deficiencies that occurs in olive trees is phosphorus deficiency. Microbes known as phosphate soluble bacteria convert the form found in soil into a usable source of phosphorus for the olive tree. The bacteria cause an increase in carbon dioxide, which somehow stimulates the phosphorus to change forms. The biofertilizer that contains phosphorus appears to work better in conjunction with an application of ascorbic acid. Together the two increase flowering, fruit yield and fruit size.

Olive Waste as Biofertilizers

Olive oil is made by pressing the fruit from the olive tree to extract the oil. It leaves behind a residue called olive pomace and wastewater from the production. The waste products have a nitrogen-fixing bacteria, Azobacter vinelandii, introduced. The bacteria removes the toxins and turns the waste into a biofertilizer. The end result is used as a soil conditioner and liquid fertilizer, which may be applied back into the olive orchard. It only takes three to seven days to achieve a 96 per cent removal of toxins and turn the waste into a beneficial product.

INTEGRATION OF ORGANIC FARMING

Rice and fish are the staples for the people in most Asian countries. Over 90 per cent of the world's area under rice, equivalent to approximately 134 million hectares, is grown under flooded conditions providing not only home to a wide range of aquatic organisms, but also offering opportunities for their enhancement and culture. Asia is the largest producer of rice (91%) with an average productivity of 3.9 t/ha. India has the maximum rice area, 44.5 Mha, although China with second largest area under rice, 30.5 Mha, is the world's largest producer of rice, with average productivity of over 6 t/ha. However, the productivity of the irrigated rice ecosystem, that accounts over 76% of the global rice production under continuous and intensive cropping conditions is either stagnating or declining.

Externalities of Green Revolution

With the introduction of high yielding technology although rice yield has increased substantially in irrigated rice tracts in India, there has been no significant increase in yield under rain fed and lowland rice ecologies that account for over 50 per cent of rice area. Consequently, in states like Assam, Bihar, Orissa and Madhya Pradesh and Kerala the realised yield have been very low as compared to irrigated ecologies.

There is a strong view that the Green Revolution paradigm adopted in these places in tune with the National agenda has been inappropriate owing to the complete neglect of the natural situations. The technology supposed to be scale and resource neutral has therefore been confined to the regions with favourable farming situation.

It is argued that in states likn Kerala, the increase in productivity with the advent of high yielding technologies was not commensurate as compared to escalation in cost of production. This has been partly attributed to the high rainfall situations, undulating topography, and water logged rice-growing situations in high rainfall tropics at variance from that of the semi arid irrigated areas.

These unique situations call for technologies tailored to the specific environmental conditions.

Some of the second generation problems in regions of high productivity have been identified as:

- Declining or plateauing of farm productivity due to depletion of organic matter, coupled with depletion of soil fertility due to over mining of native nutrient reserve.
- Declining fertilizer use efficiency, groundwater depletion due to overexploitation in utter disregard to level of natural recharge ability.
- *Increasing problems of salinity*: Alkalinity in the command areas due to excessive and indiscriminate use of irrigation water and
- *Build up of disease*: Pest pressure by continuous cropping and increased varietal uniformity and excessive dependence of on indiscriminate use of pesticides that eventual emergence and resurgence of pests thereby increased cost on plant protection.

Lowland Rice Ecologies

Endowed with high rainfall, undulating topography, and water logged situations, the rice growing situations in high rainfall tropics are at variance from that of the semi arid irrigated areas. These unique situations call for technologies tailored to the specific environmental conditions in fragile rice ecologies. Over centuries, our forefathers have therefore evolved a farming strategy in consonance with the rigid environmental conditions such as saline to acid soil and rising flood conditions in these places. This calls for a shift to an ideal cropping/ farming system that ensure high productivity, reasonable economic returns and least risk to natural endowments. It has been highlighted that the high yielding technologies were highly productive and profitable in rotation with varied crops and cropping sequences.

BENEFITS OF ORGANIC FARMING

Organic agricultural practices are based on a maximum harmonious relationship with nature aiming at the non-destruction of the environment. The developed nations of the world are concerned about the spreading contamination of poisonous chemicals in food, feed, fodder and fibre.

Naturally, organic farming system is looked upon as one of the means to remedy these maladies there. However, the major problem in India is the poor productivity of our soils because of the low level content of the organic matter.

The efficiency of the organic inputs in the promotion of productivity depends on the organic contents of the soil. There were many resemblances of organic farming principles in the traditional agriculture of India. But the former gives a more open and verifiable scientific foundation than the latter.

Healthy Foods

A study conducted in USA on the nutritional values of both organic and conventional foods found that consumption of the former is healthier. Apples, pears, potatoes, corn, wheat and baby foods were analysed to find out 'bad' elements such as aluminum, cadimum, lead and mercury and also 'good' elements like boron, calcium, iron, magnesium sellenium and zinc. The organic food, in general, had more than 20 per cent less of the bad elements and about 100 per cent more of the good elements.

Improvement in Soil Quality

Soil quality is the foundation on which organic farming is based. Efforts are directed to build and maintain the soil fertility through the farming practices. Multicropping, crop rotations, organic manures and pesticides, and minimum tillage are the methods employed for the purpose. Natural plant nutrients from green manures, farmyard manures, composts and plant residues build organic content in the soil. It is reported that soil under organic farming conditions had lower bulk density, higher water holding capacity, higher microbial biomass carbon and nitrogen and higher soil respiration activities compared to the conventional farms. This indicates that sufficiently higher amounts of nutrients are made available to the crops due to enhanced microbial activity under organic farming. The effect of organic cultivation on soil fertility as reported at the farm of Central Institute for Cotton Research, Nagpur is given in Table.

Increased Crop Productivity and Income

Field trials of organic cotton at Nagpur revealed that during the conversion period, cotton yield was low compared to the conventional and integrated crop management. However, the yields of organic cotton started rising from third year.

Cotton yields under organic, conventional and the mixed systems were 898, 623 and 710 kg/ha respectively at the end of the fourth year of the cultivation. The yield of soyabean under organic farming was also the highest compared to the other two systems. The Central Institute for Cotton Research, Nagpur conducted a study of economics of cotton cultivation in Yavatmal district of Maharashta. The cost of cultivation of cotton was lower in the organic farming than in the modern system. The low costs were due to the non-use of fertilizers and chemical insecticides.

As a result of the low yields during the conversion period, the net income from the organic farm was lesser than the conventional farm. But the yield under organic method increased progressively equalling it to that of the conventional system by the sixth year. The input costs were low under organic farming and with a 20 per cent of premium prices of output, the net income increased progressively from fourth year under organic fanning. The appreciation of net income from organic cotton cultivation by the sixth year was 80 per cent over the conventional crop.

Results reported from 1050 field demonstration cum trials under the National Project on Development and Use of Biofertilizers in different parts of the country show* an increase of 4 per cent in yield in plantation crops, 7 per cent in fruit crops, 9 per cent in wheat and sugarcane, 10 per cent in millet and vegetable, 11 per cent in fibre, condiments and spice crops, 14 per cent in oilseeds and flowers and 15 per cent in tobacco.

A study of 100 farmers in Himachal Pradesh during a period of 3 years found that the total cost of production of maize and wheat was lower under organic farming and the net income was 2 to 3 times higher. Both productivity and premium prices contributed to the increased profitability. Another study of 100 farmers of organic and conventional methods in five districts of Karnataka indicated that the cost of organic farming was lower by 80 per cent than that of the conventional one.

Table. The Cost Benefit Ratios Described for Various Crops

Crops	Organic	Inorganic
Groundnut	1:1.26	1:1.31
Jowar	1:1.36	1:1.28
Cotton	1:1.34	1:1.24
Coconut	1:1.70	1:1.31
Banana	1:3.66	1:2.82

Low Incidence of Pests

The study of the effectiveness of organic cotton cultivation on pests at the farm of Central Institute for Cotton Research, Nagpur revealed that the mean monthly counts of eggs, larva and adults of American BoUworm were far lesser under organic farming than under the conventional method.

Bio-control methods like the neem based pesticides to Ti-ichoderma are available in the country. Indigenous technological products such as Panchagavya which was experimented at the University of Agricultural Sciences, Bangalore found to control effectively wilt disease in tomato.

Employment Opportunities

Organic farming requires more labour input than the conventional farming system. Thus, India which has a very large amount of labour unemployment and under employment will find organic farming an attraction.

Moreover, the problem of periodical unemployment will also get mitigated because of the diversification of the crops with their different planting and harvesting schedules resulting in the requirement of a relatively high labour input.

Indirect Benefits

Several indirect benefits from organic farming are available to both the farmers and consumers. While the consumers get healthy foods with better palatability and taste and nutritive values, the farmers are indirectly benefited from healthy soils and farm production environment. Eco-tourism is increasingly becoming popular and organic farms have turned into such favourite spots in countries like Italy. Protection of the ecosystem, flora, fauna and increased biodiversity and the resulting benefits to all human and living things are great advantages of organic farming which are yet to be properly accounted for.

Proposed Objectives

The broad objectives of organic farming in India can be the following in the light of the discussions on the adverse effects of the conventional farming system the country practiced for about 30-40 years and the potential benefits of the organic methods:

- Sustainable agriculture
- Increasing agriculture production
- Food self-sufficiency
- Environmental protection
- Conservation of natural resources
- Rural development.

RELEVANCE OF ORGANIC FARMING

The relevance and need for an eco-friendly alternative farming system arose from the ill effects of the chemical farming practices adopted w^orldwide during the second half of the last century. The methods of farming evolved and adopted by our forefathers for centuries were less injurious to the environment. People began to think of various alternative farming systems based on the protection of environment which in turn would increase the welfare of the humankind by various ways like clean and healthy foods, an ecology which is condusive to the survival of all the living and non-living things, low use of the non-renewable energy sources, etc. Many systems of farming came out of the efforts of many experts and laymen. However, organic farming is considered to be the best among all of them because of its scientific approach and wider acceptance all over the world.

The International Scene

The negative effects of modem chemical based farming system were first experienced by those countries, which introduced it initially. So, naturally, it was in those countries organic farming was adopted in relatively large scales. There are very large organisations promoting the organic farming movement in European countries, America and Australia etc. These organisations the International Federation of Organic Agriculture Movements and Greenpeace have studied the problems of the chemical farming methods and compared the benefits accruing to the organic faming with the former. Organic farming movements have since spread to Asia and Africa too.

IFOAM was founded in France in 1972. It spearheads and coordinates organic farming efforts the world over by promoting organic agriculture as an environment friendly and sustaining method. It focuses on organic farming by highlighting the minimum pollution and low use of non-renewable natural resources through this method. It has about 600 organisational members spread over about 120 countries including India.

IFOAM undertakes a wide range of activities related to organic farming such as exchanging knowledge and thoughts among its members; representation of the movement in governmental, administrative and policy making forums in the national and international arena; updating of production, processing and trading standards; formulation and coordination of research projects; and holding of international conferences and seminars. IFOAM participates in the activities related to organic farming under the auspices of the United Nations and keeps active contacts with several international NGOs. The Food and Agriculture Organisation of the United Nations provides support to organic farming in the member countries. It also attempts the harmonization of national organic standards, which is absolutely essential to increase international trade in organic products. The FAO has, in association with the World Health Organisation, evolved the Codex Alimentarius for organic products.

Organic farming has several advantages over the conventional one apart from the protection of both the environment and human health. Improved soil fertility, better water quality, prevention of soil erosion, generation of rural employment, etc. are some of them.

The concept of quality food has undergone a drastic change over the past few decades. It does give emphasis on the characteristics of the end product, but the process and method of production and transport are now considered equally important. Not only the importers but also the domestic retailers have their own quality specifications or standards, which in many cases are tougher than those of the government regulations.

Consumers have become health conscious and are willing to pay for the clean, healthy and natural food. Many developed countries have various support programmes to help organic farming with financial incentives and technical guidance. The organic food market in the world has grown rapidly in the past decade. International trade in organic foods showed an annual growth rate of about 20-22 per cent during this period.

Many retail chains and supermarkets in advanced countries are accorded with'green status' to sell organic foods. The organic food processing industry is considered nature friendly and thus encouraged.

Table presents the size of the organic food markets in some developed countries in 2000. The US food market tops with a value of $8.0 billion, followed by Germany and Japan with $2.5 billion each while France accounts for $1.25 billion. Moreover, the organic foods command higher prices which often range from 10 to 100 per cent more than the conventional ones.

The important organic products traded in the international market are dried fruits and nuts, processed fruits and vegetables, cocoa, spices, herbs, oil crops and derived products, sweeteners, dried leguminous products, meat, dairy products, alcoholic beverages, processed foods and fruit preparations. Cotton, cut flowers, animals and pot plants are major non-food products in the world markets.

Organic farming is bound to grow around the world as many countries are developing their own standards and regulations. The US and the European Union have very comprehensive National Organic Programmes and the early nineties have seen organic farming regulations in Japan, Canada and Australia. New Zealand, Israel and Brazil have adopted the organic standards equivalent to those of USA and the European Union. China, Thailand, South Korea, Philippines, Turkey and Mexico have established certifying agencies. India too has adopted the National Programme for Organic Production with national standards.

There were about 35,000 certified organic farms in 17 countries during 1998 covering an area of more than a million hectares. A survey held recently revealed that 6-7 per cent farms in USA and Germany are certified organic farms.

Many organic inspection and certifying agencies like Crop Improvement Association, IFOAM, Ecocert, BCS, SKAL. KRAV, IMO, FVO, QAI, Oregon, Tilth and NASAA have strong international presence and have even developed the ability to assist marketing of organic products world wide.

Certification of products is the foundation on which organic agriculture rests. Its role in production and marketing of organic foods becomes a significant corner stone in the creation of a confident consumer community.

The need for a trustworthy and healthy organic market is important as the consumers are becoming increasingly separated from the food production process.

The void is filled in by a plethora of processing, distribution and marketing intermediaries. The consumers should possess knowledge and confidence that the products they buy are truly organic. Such environment can be created by an organised system of inspection and certification supported by law and regulations of the state. Organic products enjoy trust based on the label or logo approved by the certification process.

The aim of organic certification is, therefore, to:

- Provide an identifiable label (a logo or a symbol); and
- Give an assurance to the consumers that the product is truly organic.

It is widely recognised the world over that the certification of organic products should be based on the following principles:

- Organic production and processing standards should be clearly laid down.
- The conformation of production and processing to these stands must be verified.
- Organic labels should be permitted only to those produces, which are found conforming to the set standards.

Thus, a label on an organic product conveys that the manufacturer has a license for organic production, an independent agency has inspected the production/processing practices followed by the producer, and compliance of the set standards is made.

Such an assurance becomes crucial in the generation of consumer confidence in the organic products, particularly as they are costlier than the conventional ones. The marketing agents of organic products also need such an assurance to maintain fair trade practices essential to the growth and sustainability of the demand for such products.

Growth of Organic Farming

Organic farming has spread to about 100 countries around the world. An estimate in 2004 puts about an area of 24 million hectares under oi'ganic farming worldwide. Australia with its 10.5 million hectares leads the countries. However, much of this area is pastoral land for grazing. Argentina with 3.19 million hectares, Italy with 1.83 million hectares and USA with 0.95 million hectares follow suit.

The importance of organic farming is growing in many countries.

Austria and Switzerland have about 10 per cent of their food system under organic agriculture. The annual growth of organic farming is estimated to be about 20 per cent in USA, France, Japan and Singapore.

The Developed World

Table shows the top 15 countries in terms of area under organic farming, percentage of organic area to the cultivated area the number of organic farms: Australia had the largest area under organic farming and Argentina was the next, followed by Italy. However, in terms of percentage of organic farming area to that of the cultivated, Australia was the fourteenth and Italy fourth. In terms of number of farms under organic agriculture, Italy topped. An examination of all these three characteristics, indicates that Italy, Germany and Austria are the leading countries in organic agriculture and they are among the developed nations of the world. Not even a single developing country except Argentina in terms of area found a position among the top fifteen described in the Table. However, it should be remembered that the US was the fourth in terms of area under organic farming and had the largest market for organic products in the world.

USA

Organic farming did spread fast in USA during the 1990s. The growth was in response to the rising demand for organic products both in the national and international markets. The US agriculture farms view organic farming as a method to reduce input costs and use of non-renewable resources, which increases accessibility to the high value markets and assures premium prices for the produces and above all increased farm income. The US organic farming system uses eco-friendly agriculture practices in cultural and pest management and avoids the application of chemicals. Antibiotics and hormones are not used in livestock production.

The growth of consumer demand for organic products has also resulted in the specialization in growing, processing and marketing of these products. The implementation of national organic standards in 2002 by the USDA has further boosted the demand for organic products in the country.

Because of increasing prices and ineffectiveness of chemicals, many farms in USA do not favour intensive chemical based agi-iculture. A survey of certified organic farms revealed that more than 60 per cent of them grew vegetables and they had livestock for manure to use in the farm. The farmers had identified three problem areas, which required study.

They were:

1. The need for increasing demand for organic products,
2. The influence of cultural practices on quality and nutritional value of the products; and
3. The relationship between plant nutrition and resistance to pests.

In another study of consumer demand for organic foods, it was found that about 30 per cent of shoppers used some organic food in a year, while 10 per cent used it once a week. The value of sales of organic products in US was

$1.5 billion in 1992 and $8 billion in 2000. Many methods to promote sales and consumption of organic food have been introduced. The US Senate approved a bill to make a part of the Federal School Lunch Programme organic. Organic food shows are held annually to give awards for the best organic foods and drinks. The Federal government encourages the planting of native plants for Municipal planting programmes. The government has initiated polices to reduce the application of pesticides and labelling food to indicate the amount of pesticides used to produce the commodity is compulsory. There are about 17 organic certification agencies in US.

The certification and labelling laws are very rigid in United States. Imports of organic products must conform to the standards laid down by the USDA and naturally, the ability of many developing countries to comply with these requirements is limited. A grower or a dealer in need of certification has to lodge a plan to an accredited agency of the USDA. There are both private and state certification programmes. The plan must contain all details in relation to the farming or handling methods and the materials those will have to be used therein. The anticipated improvements and future aims should also be a part of the plan. The farmers or dealers of wild crops should also submit a plan mentioning that the harvesting techniques to be employed will not cause any damage both to the produce and the environment.

All records related to the management practices and the materials utilized in organic production should be maintained for five years. Certification will be approved only if the crops are grown on the land free of prohibited substances for three years prior to harvest. The period of conversion is three years between the conventional and organic farming and the crops grown during this period cannot be labelled organic or any other label indicating that they are under conversion.

The Organic Foods Production Act contains a wide and detailed coverage of organic farming methods and materials along with the soil fertility, application of manure, crop rotation and composting. The National Organic Standards Board's recommended ingredients include crop residues, crop wastes from food processing operations, animal manure, yard wastes from private and municipal sources, or other vegetable by-products. Municipal solid waste compost and sewage sludge composts and the use of prohibited material as composts/ingredients are banned. There is a National List of acceptable and banned materials which includes pest control treatments and other agricultural inputs like fertilizers and seed treatments. An evaluation of all agricultural inputs has been made for their long term effect on the environment irrespective of the fact that they are synthetic or natural.

The exporters of other countries to the US market will have to face these very tough specifications, particularly if their national standards are not equivalent to those of the USDA.

Italy

The European Union of 25 countries had 5.5 million hectares under organic farming at the end of 2002 managed by 1,60,000 organic farms. This constitutes about 3.4 per cent of the agricultural area and 1.7 per cent of farms in EU. It is estimated that the growth in the area was 9 per cent over the last year. The increases came mainly from France, Spain and the United Kingdom. The European Commission has formulated the 'European Action Plan for Organic Food and Farming' on 10* June 2004. It proposes 21 specific actions to promote organic farming in the EU countries and the Action Plan is with the European Parliament for consideration. As said earlier, the progress achieved by Italy, Austria and Germany was impressive and as such it is discussed here.

The pioneers of organic farming in Italy had contributed immensely to the spread of organic agriculture, which is acclaimed to be one of the best in the world. Farmers, medical doctors, food processors, traders and consumers were in the forefront working hard from the 1950s for the cause of organic farming in the country.

Growth of organic farming in Italy has been spectacular during the period of 1990-2000. The organically cultivated area increased from 13,000 ha to more than a million ha which came to about 7.2 per cent of the total cultivated area. The number of farmers was about 13,00 in 1990, which shot up to 47,357 at the end of 2000. At the end of 2002, the area increased to 1,83,000 ha and the number of farms to 56,440. The reasons for such a substantial increase were the food scandals like the mad cow disease leading to the search for safe foods, the constant endeavor for alternative farming methods which were environmentally, technically and economically attractive to the farmers and the liberal subsidies.

The farmers who opted for organic methods prior to 1990 were people of non-agriculture background, people searching for novel life styles, etc. A significant proportion of them consisted of non-Italians like, Germans, Swiss, Austrians and Americans. They had the unity of purpose in establishing cultural associations in cooperation with like minded people by setting up marketing links with organic farms. However, the decade of 1990-2000 saw that the new entrants into organic farming were Italians and many of them did not have experience of the new methods. But they possessed a good knowledge of the ecology and understood the need for a marketing system.

The period also witnessed a change in the geographical distribution of area under organic cultivation. Sardinia became an important region in organic farming with the extension of subsidies to pasture lands used for sheep rearing and milk production. Sardina and Sicily account for about 44 per cent organic area in Italy.

At present 41 per cent of the organic area is under pastures. Cereal crops are raised on 18.2 per cent of the area and the olive cultivation alone has the highest single crop share of 10 per cent. The high growth of organic production led to the growth of the processing units to the extent of 142 per cent during 1997-2000 alone. Presently, there are about 42,000 processing units of which 1,300 are established by the farmers themselves. Organic farming has resulted in the setting up, conversion or expansion of new economic activities in the rural areas of Italy.

The first organic shop had come up in Milan, the industrial and financial capital of Italy, during the 1970s under a consumers' cooperative venture. The most important fact of the market distribution of organic products in Italy is the predominant position of the farmers' or consumers' cooperatives. The distribution was mostly done during the immediate post 1970s by small grocery stores and the major share was that of the direct sales in the open market places and at the farms. After 1990, specialized markets, supermarkets, franchisees, school canteens, etc. became the major channels. Supermarkets are entering the field in large numbers since 1999 with their own labels to convince the consumer of the reliability of the products in the midst of numerous labels. Diary products have a share of 26 per cent of the consumer expenditure on organic products; fruits and vegetables accounts for 13 per cent; breads and biscuits had a share of 12 per cent. A growing number of kindergartens, school and university canteens, restaurants and factory Ccinteens are opting for organic foods in the country. A survey of the motives of the consumers of organic foods revealed that every one was going for organic products to protect the health of his/her family.

The support for organic farming by the European Union by various legislations providing subsidies was instruinental in the growth of organic farming in Italy. The devolution of power to the Regional Government on agriculture policy in 1972 has resulted in different levels of subsidies in different regions/crops. There are nine organic certification bodies in Italy. Some of them were earlier cultural associations promoting organic farming. A single agency named ICEA has a share of 24 per cent of the certified farms coming to about 29 per cent of the area. These nine agencies have about 900 local offices, 1000 staff and technicians for inspecting farms, processing units and storages for giving certifications. Universities and Public Research Centres are becoming increasingly interested in research and extension activities in relation to organic agriculture. A National School of Organic Farming has been established by the Ministry of Agriculture to impart education.

Austria

Austria is a leading country in Europe in the adoption of organic farming. At the end of 2003, about 11.6 per cent of the cultivated area which came to 2,96,154 ha was under organic agriculture. The number of organic farms came to 18,576, which was 9.2 per cent of the total number of agriculture farms in Austria. The conversion of cultivated land to organic method of agriculture was slow till 1990. The federal government introduced subsidies for organic conversion of cultivation in 1991. The government programme of agri-environment of 1995 and the conversion of grass lands into organic farms gave a great fillip to the organic movement. There was a decrease in the conversions in 2000, but it increased substantially from 2001 to 2003. The decline was due to the difficulties in marketing the organic produces in the country. This was an indication that the emphasis on farm production was not enough to promote organicagriculture.

The increasing phase from 2001 to 2003 saw about 700 farms having an area of 30,000 ha being converted. This came from lands cultivating cereals and vegetables. One of the reasons for the change is attributed to the realization of the benefits of organic farming by the young farmers with a better concern for ecology and understanding of economics than their parents. The organic farmers belonged to a number of associations and the latter had different approaches/objectives, regional differences and marketing techniques. About 64.5 per cent of the organic farmers were members of these associations, which functioned under two umbrella organizations, named ARGE and OelG. These associations have their own production and quality standards, which in many cases are more stringent, than those of the State. Presently there are 11 major organic farming associations in Austria and while 6 of them come under ARGE, the remaining five and 4 marketing associations are members of OelG, the latter being the one giving importance of marketing too in the promotion of organic agriculture in Austria.

The concentration of organic farms is in the regions with large areas of grass land. It is easier for the conversion of the grass land to organic farming as they used to be cultivated extensively. Grass land farms constitute about 66 per cent of the total organic farms. The average size of an organic farm in Austria is 16 ha.

Certification of organic farms in Austria is done with the provincial government and a contract between the farm and an inspection agency is essential for registration. An approved organic label is available to guarantee the quality of the organic products. The conditions to be fulfilled by an inspection agency for accreditation are very tough in Austria. 17,500 registered organic farms receive government subsidies and about 50 per cent of it is born by the European Union and balance by the federal and provincial governments.

The market value of Austrian organic products was 2.18 billion Euro in 2000 which was 3 per cent of the total food market. About 70 per cent of these products is distributed by the conventional food chains. Now, the distribution of organic foods through the supermarket chains is gaining momentum. It is reported that the supermarkets selling organic products enjoy an improved image among their customers. Organic farms' cooperatives also market organic products. Cooperative ventures between the organic farmers and hospitals, nursing homes, etc. have come up to sell organic products. Home delivery services and organic shops are also established, particularly in large cities. Agricultural Schools and Colleges offer training programmes/courses in organic farming and promote scientific research. The Institute for Organic Farming at the University of Vienna is a major centre for research in organic agriculture.

Developmient of a strategy for co-operation among the organic farmers, their associations, processors and traders is considered a major challenge in the promotion of organic farming in Austria.

Germany

Germany, where the IFOAM is located at Tholey-Theley since 1987, is a leading nation in the adoption of organic farming in Europe. The IFOAM has its Regional Group founded in 1991 with the German speaking nations as members. Germany has about 3 per cent of its total agricultural area under organic farming against 2 per cent of the European countries as a whole. The country has also a large number of processors and importers of organic produces. So, it would be useful to examine the organic farming movement in that country in some detail.

The history and development of organic agriculture in Germany begins from the days of bio-dynamic farming initiated in 1924. Bio-dynamic theory of farming treats farm as a living being which responds to non-material influences, called dynamic forces. The proponents of bio-dynamic farming believe that these forces can be enhanced by bio-dynamic preparations. Even certification for bio-dynamic agriculture started in 1954 in the country. Organic-bio-dynamic farming was developed in Switzerland in the early twentieth century. Hans Mueller and his wife Maria Mueller were the spirits behind this movement in Switzerland. Mueller had developed the method during the 1950s and the theoretical basis was provided by Hans-Peter Rusch, the German microbiologist and medical doctor. The damage inflicted to the environment by the modern farming system was clear by the end of 1960s. The ill effects of pollution were obvious and a producer organisation, named Bioland was established in 1971 in Germany. Another organisation for the organic cultivation of fruits and vegetables had been founded in 1961. A federation for organic farming associations was established in 1988 consisting of six producer organisations. A common basic standard for organic agriculture was laid down in 1984 as a framework for individual organization.

Organic farming spread very fast in Germany during the subsequent period. State funding for organic agriculture was available from 1989. The unification of both the Germanys in 1990 saw fast growth of organic farming in the former East German areas. The trade sector felt the absence of a German logo for organic products and the seal 'Biosiegel' was launched by the federal government. The German government has put the goal of bringing 20 per cent of the agricultural land under organic farming by 2010 from the 3 per cent of 2002. A series of measures were introduced to achieve the goal. These steps are under the EU programme of support to organic agriculture and the federal programme for organic farming.

There are eight organic producer organisations in Germany and they represent about 60 per cent of the organic farming community. All these associations have the common organic seals for labelling the certified products, both primary and processed. The organic producers, processors and traders of the countiy have formed a common organization aiming at the propagation of achievements of organic production, protection of environment etc. There are several other organisations functioning in relation to organic farming in Germany. SOL, FiBL, Schweisl\irh Foundation, Foundation Future Agriculture and Society for Ecological Animal Husbandry are some of the institutions contributing to the spread of organic farming in Germany.

Organic farming has gained acceptance in the former East Germany compared to the West. The reasons are the substantial changes in the German agriculture after the reunification, existence of large conservation areas in which organic farming can well be fitted into. Large areas in the eastern part are designated as disadvantaged and incentives are provided there to go for organic farming. The arguments for and against organic farming between the organic and conventional farmers are intense in the western parts of Germany and this has to some extent slowed down the spread of the organic agriculture there. However, government regulations and the support of European Union Regulations have made the organic agriculture accepted in the country.

The standards and certification process for organic farming are stricter in Germany than in other European Union countries. The German regulations restrict certification to the whole farm and the processing standards are tough. There are private and state standards and the inspections are done according to both of them. The use of genetically engineered organisms and techniques are banned in organic agriculture. There are more than 20 inspection bodies and the 22 states of the country control them. Table gives the estimated organic production in Germany. The German organic food market was worth about 3 billion Euro in 2000. Both the federal and state governments extend financial support to organic farming. Farmers are encouraged to adopt organic methods from 1989 and those who adopt production methods compatible with the protection of environment are given assistance. In addition to the direct subsidies, marketing assistance is also provided to the organic farms. Subsidies for producers' marketing organisations, processing and market promotion are available.

Advisory services are provided to the farmers for organic farming. Advisory services were organised by the farmers themselves during the initial period of the introduction of organic farming in Germany. Now producer associations with governmental financial support offer such services. Farmers themselves hire advisors with state support and official advisors are also available. Research and training facilities are existing in almost all the German agricultural universities and technical colleges in organic farming. Degree courses are offered in organic agriculture and organic farming research is also funded by the state. A state research institute for organic farming was established in 2000.

IMPORTANT ROLE OF BIOFERTILIZER IN CROP PRODUCTION

Biofertilizers have an important role to play in improving the nutrient supplies and their crop-availability in upland crop production. Although Rhizobium is the most researched and well known among these, there are a number of microbial inoculants with possible practical application in upland crops where they can serve as useful components of integrated plant nutrient supply systems. Such inoculants may help in increasing crop productivity by way of increased biological nitrogen fixation (BNF), increased availability or uptake of nutrients through solubilization or increased absorption, stimulation of plant growth through hormonal action or antibiosis, or by decomposition of organic residues.

The scope of this chapter is confined to microbial inoculants of N2-fixing bacteria, vesicular arbuscular mycorrhizae (VAM), phosphate solubilizers and plant-growth promoting rhizobacteria (PGPR) and their role in production of upland crops. Aspects determining the success of inoculant technology are discussed here rather than presenting an up-to-date account of literature on the performance of biofertilizers. Rhizobial inoculants are dealt with in detail as a general example and for others, our discussion is restricted to those points which are specific to a particular group of inoculants.

Nitrogen-fixing Bacterial Inoculants

Above every hectare of land at sea level, there are 78,000 tonnes of inert nitrogen gas (N_2). Nitrogen is the most limiting nutrient for increasing crop productivity. Only a few procaryotic organisms are able to "fix" N_2 directly through a biological process. Annually BNF is estimated to be around 175 million tons of which close to 79% is accounted for by terrestrial fixation.

The N_2 fixers involved are metabolically diverse but the process is similar and depends upon:

- Nitrogenase (N_2ase) enzyme complex,
- A high energy requirement (ATP)
- Anaerobic conditions (for N2ase activity) and,
- Source of strong reductant. Amongst N_2-fixing bacteria *viz.*, Rhizobium, Azospirillum, and Azotobacter, the most widely used inoculant is Rhizobium.

Rhizobium

Symbiotic N2-fixation by Rhizobium with legumes contributes substantially to total BNF. Rhizobium inoculation is a well known agronomic practice to ensure an adequate nitrogen nutrition of legumes in place of fertilizer N.

Table. Estimates of Nitrogen Fixed by Some Legumes.

Crop	Nitrogen Fixed (kg/ha)	Crop	Nitrogen Fixed (kg/ha)
Alfalfa	100-300	Lentil	35-100
Clover	100-150	Green gram	50-55
Chickpea	26-63	Pigonpea	68-200
Cluster bean	37-196	Soybean	49-130
Cowpea	53-S5	Peas	46
Groundnut	112-152	Fenugreek	44

Classification

The genus Rhizobium consists of six distinct species based largely on the cross inoculation group concept. The assumption in this classification is that those leguminous plants falling within a particular infection group were nodulated by a particular species of nodule bacteria. More than twenty cross-inoculation groups have been established, but only seven have achieved prominence. The validity of the cross-inoculation groups has been questioned as many legumes are nodulated by rhizobia of other host-bacterial groups. With repeated evidence of anomalous cross-infection among the different plant groups, new classification has been proposed. The slow-growing rhizobia are grouped under the genus Bradyrhizobium and the fast growers under the genus Rhizobium.

The new classification is more complex as the same host may be nodulated by fast and slow-growing strains. Moreover, the agricultural significance of the cross inoculation groups still remains a key feature of the established taxonomic system. In this article we have used the classification based on cross-inoculation groups.

The cross-inoculation groups and Rhizobium-host plant associations are listed below:

Cross-inoculation	**Rhizobium Species**	**Host Genera**	**Legumes Included Group**
Alfalfa group	R. meliloti	Medicago	Alfalfa
		Melilotus	Sweet clover
		Trigonella	Fenugreek
Clover group	R. trifolii	Trifolium	Clovers
Pea group	R. leguminosarum	Pisum	Pea
		Vicia	Vetch
		Lathyrus	Sweet pea
		Lens	Lentil
Bean group	R. Phascoli	Phaseolus	Beans
Lupine group	R. lupini	Lupinus	Lupines
		Ornithopus	Serradella
Soybean group	R. japonicum	Glycine	Soybean

Cowpea group	Rhizobium sp.	Vigna	Cowpea
		Cajanus	Pigeonpea
		Cicer	Chickpea
		Lespedeza	Lespedza
		Crotolaria	Sunnhemp
		Arachis	Groundnut
		Phaseolus	Lima bean
		Pucraria	Kudz

Need for Inoculation

The most important point is do we need inoculation of legumes in a region where these crops have been grown over long periods? Development of an inoculant industry in many countries has been largely motivated by the desire to introduce legume species to new areas. Most cultivated tropical soils are assumed to have relatively large populations (> 100/g dry soil) of rhizobia capable of nodulating the legumes grown in such soils. The need to inoculate the legumes grown on cultivated soils must be assessed by considering the interacting factors between the soil, the host plant and *Rhizobium.*

Presence of nodules on plant roots does not necessarily mean that sufficient N2 is being fixed for maximum benefit to the host plant. In groundnut or pigeonpea nodulation occurs naturally at most locations due to the cross-species promiscuity of the cowpea rhizobia. However, the ability to fix high amounts of N (efficiency) is governed by the symbiotic capability between *Rhizobium* and the host plant. Hence, it may be necessary to introduce superior (more competitive and efficient) strains of *Rhizobium* to ensure adequate N2fixation for maximum growth and yield of the host plant.

In a survey of groundnut crops grown in farmers' fields in southern India, 52 out of 95 fields showed inadequate nodulation with less than 10 per cent N2-fixing (acetylene reducing) activity of what can be obtained under reasonable field conditions.

The results of surveys of farmer's gain and fodder legume crops also revealed poor nodulation in large areas and good noduation only in a few pockets. Poor nodulation in farmer's fields could be due to several lactors eg. lack of appropriate rhizobia in soil, deficiency or toxicity of a nutrient, unfavourable conditions like prolonged water logging, unsuitable pH, abundance ofbacterial predators, pests and disease attack, etc. Although adequate nodulation was observed in some parts, ineffective nodules exceeded the number of effective nodules. Out of 87 groundnutrhizobial strains isolated from different parts of India, only 5 were found to be effective.

THE BEST FEED FOR CATTLE AND POULTRY IN ANABAENA AZOLLAE

Azolla is a free floating water fern that floats in water and fixes nitrogen in association with the nitrogen fixing blue green algae, *Anabaena azollae.* Azolla is considered to be a potential biofertilizer in terms of nitrogen contribution to rice.

Long before its cultivation as a green manure, Azolla has been used as a fodder for domesticated animals such as pigs and ducks. In recent days, Azolla is very much used as a sustainable feed substitute for livestock especially dairy cattle, poultry, piggery and fish.

Azolla contains 25 – 35 per cent protein on dry weight basis and rich in essential amino acids, minerals, vitamins and carotenoids including the antioxidant b carotene.

Cholorophyll a, chlorophyll b and carotenoids are also present in Azolla, while the cyanobiont *Anabaena azollae* contains cholorophyll a, phycobiliproteins and carotenoids. The rare combination of high nutritive value and rapid biomass production make Azolla a potential and effective feed substitute for live stocks.

Inputs Required

Azolla fronds, Polythene sheet, Super phosphate and Cow dung.

Methodology

The area selected for Azolla nursery should be partially shaded. The convenient size for Azolla is 10 feet length, 2 feet breadth and 1 feet depth. The nursery plot is spread with a polythene sheet at the bottom to prevent water loss. Soil is applied to a depth of 2 cm and a gram of super phosphate is applied along with 2 kg of vermicompost or cow dung in the nursery for quick growth. Azolla mother inoculum is introduced @ 5 kg/plot.

The contents in the plot are stirred daily so that the nutrients in the soil dissolve in water for easy uptake by Azolla. Azolla is harvested fifteen days after inoculation at the rate of 50-80 kg/plot. One third of Azolla should be left in the plot for further multiplication. Five kg cow dung slurry should be sprinkled in the Azolla nursery at ten days intervals. Neem oil can be sprayed over the Azolla at 0.5 5 level to avoid pest incidence.

Animal	Dosage / Day
Adult cow , Buffalo, Bullock	1.5-2 kg
Layer, Broiler birds	20 – 30 grams
Goat	300 – 500 grams
Pig	1.5 – 2.0 kg
Rabbit	100 gram

Value of the Technology

The egg yield is increased in layer birds due to Azolla feeding. The Azolla fed birds register an overall egg productivity of 89.0 per cent as against 83.7 per cent recorded by the birds fed with only concentrated feed.

The average daily intake of concentrated feed is considerably low (106.0 g) for birds due to Azolla substitution as against 122.0 g in the control birds. More impotantly Azolla feeding shows considerable amount of savings in the consumption of concentrated feed (13.0%) leading to reduced operational cost.

By considering the average cost of the concentrated feed as ₹. 17/ Kg, a 13.0% saving in the consumption ultimately leads to a feed cost savings of 10.0 paise /day/ bird and hence a layer unit maintaining 10,000 birds could cut down its expense towards feed to a tune of ₹.1000/day.

Benefits

The Azolla feeding to layer birds increase egg weight, albumin, globulin and carotene contents.

The total protein content of the eggs laid by the Azolla fed birds is high and the total carotene content of Azolla eggs(440 g 100 g-1 of edible portion)is also higher than the control.

The rapid biomass production due to the high relative growth rate, increased protein and carotene contents and good digestability of the Azolla hybrid Rong ping favour its use as an effective feed supplement to poultry birds.

Table. Effect of Azolla Hybrid Rong Ping on the Nutritional Value of Egg

Parameters	Azolla Egg	Control	Percentage Increase over Control
Egg weight (g)	61.20	57.40	6.62
Albumin (g/100 g of edible portion)	3.9	3.4	14.70
Globulin (g/100 g of edible portion)	10.1	9.5	6.31
Total protein (g/ 100 g of edible portion)	14.0	12.9	8.52
Carotenes (μg / 100 g of edible portion)	440	405	8.64

Application

In Indian conditions, agriculture is very much coupled with poultry farming. Azolla is an important low cost input, which plays a vital role in improving soil quantity in sustainable rice farming. The twin potentials as biofertilizer and animal feed make the water fern Azolla as an effective input to both the vital components of integrated farming, agricultural and animalo husbandry.

Limitation

Azolla is a water fern and requires a growth temperature of 35–38º C. The multiplication of Azolla is affected under elevated temperature. Hence adopting this technology in dry zones where the temperature exceeds 40ºc is difficult.

Achievements

Azolla hybrid Rong ping had been selected to supply to the tribal population. Azolla mother inoculum nursery was laid out in villages with the help of Krishi Vigyan Kendra, TNAU, Coimbatore and Krishi Vigyan Kendra, Karamadai, women entrepreneurs were selected and one day training was imparted to them on the cultivation of Azolla. Wet biomass (Starter inoculm) were supplied at free of cost @ 10 kg/women entrepreneur during the training so as to enable them to initiate commercial Azolla cultivation in their backyards.

Azolla multiplication plots had been laid out in Narasipuram. Azolla mass production training was conducted to the SHG in Narasipuram village with the help of Kalaimagal Arts and Science College, Narasipuram, Sappanimadai (tribal village) and Avinashilingam KVK, Karamadai. With the help of Avinashilingam KVK, Karamadai Azolla trainings were conducted to women volunteers and we have established Azolla village in Karamadai. The Avin milk producers union Coimbatore and the poultry owners association, Namakkal have been contacted and explained the importance of Azolla as feed supplement.

CAUSE AND EFFECT OF RHIZOBIUM

Rhizobium is a gram negative, aerobic, rod-shaped bacterium. It contains a refractive granule. It is a soil bacterium present in large numbers in rhizosphere of legume roots.

Rhizobium invades roots of legumes and forms nodules on the roots. Inside the root nodules, the bacteria exist in various pleomorphic forms called bacterioids. The bacterioids fix the atmospheric nitrogen into ammonia. They provide the fixed nitrogen for plant's use and draw nourishments from the root cells. This type of association is called symbiosis.

Different species of Rhizobium can fix 50-200 kg nitrogen/ha/year in leguminous crops. Therefore, they have been recommended as nitrogen biofertilizers in agriculture. Different species of Rhizobium show a great degree of host specificity.

Hence they can be used as biofertilizers only for the specific crops:

- *R. meliloti* (*Medic-Rhizobium*): Lucerne and Fenugreek.
- *R. trifolii* (*Clover-Rhizbium*): Egyptian clover and clover.
- *R. Leguminosarum (Pea-Rhizobium)*: Lentil, pea, khesri (Lathyrus) and vetch.
- *R. Phaseoli* (*Bean-Rhizobium*): Bean, kidney bean and French bean,
- *R. lupini* (*Lupin-Rhizobium*): Lupines and white lupines.
- *R. japonicum* (*Soybean-Rhizobium*): Soyabean.
- *Rhizobium* sps. *(Chickpea-Rhizobium)*: Chickpea and subabul.
- *Rhizobium* sps. (*Cowpea-Rhizobium*): Sunnhemp, cluster bean, peanut, jack bean, lablab, horsegram, moth bean, green gram, blackgram and pigeonpea.

Production of Rhizobium Inoculants

Production of Rhizobium inoculant involves:

- Isolation of Rhizobium
- Identification of Rhizobium
- Establishment of starter culture
- Mass culture
- Mixing with carrier
- Packaging and storage.

1. *Isolation of Rhizobium*: Rhizobium occurs in the soil as well as in the root nodules of selective legumes. Rhizobium in the root nodules has least contaminants. Therefore, it is isolated from root nodules of a proper leguminous plant.

   The leguminous plant is carefully uprooted from the soil and the root system is washed with running water to remove the soil particles. Firm, undamaged, pink-coloured root nodules are selected visually and excised from the roots.

   The root nodules are kept immersed in 0.1% potassium chloride solution or in 0.1% acidified mercuric chloride solution for 5 minutes to sterilize the surface of the nodules.

The sterilized root nodules are then washed 5 or 6 times with distilled water. They are once again sterilized by immersing them in 90% ethyl alcohol for 10 seconds and washed repeatedly with distilled water.

The root nodules are crushed gently in a small amount of distilled water using a pestle and mortar to get a suspension.

The suspension is diluted and inoculated onto YEMA (Yeast extract mannitol agar) medium in Petri dishes. The culture plates are incubated at 28°C for about 10 days. Rhizobial cells form gummy colonies on the medium.

Table. Components of YEMA Medium

K_2HPO_4	-0.5 g
$MgSO_4.7H_2O$	-0.2g
NaCl	-0.1 g
Mannitol	-10.0g
Yeast extract	-1.0 g
Agar	-20.0 g
Distilled water	-1000 ml.

2. *Identification of Rhizobium:* YEMA medium is suitable for the growth of Rhizobium as well as Agrobacterium. Rhizobial colonies are identified from the culture in the following methods:
   - *CRYEMA Test:* A 2.5 ml of congo red dye is mixed with a litre of YEMA medium to prepare CRYEMA medium. Bacterial colonies on the YEMA medium are streaked on the CRYEMA medium and the petri dishes are incubated at 28 ± 2°C for 5-7 days.

     Rhizobial cells uptake congo red very weakly so they 27-BT form white, circular, entire, raised, convex colonies. Agrobacterium colonies, if any, look like Rhizobial colonies, but show characteristic colour of congo red. The white colonies are picked up to produce Rhizobium inoculant.
   - Microscopic Observation: β-polyhydroxybutyrate granule are picked up to establish Rhizobium inoculant. β-polyhydroxybutyrate granule in the Rhizobium. The cells of those colonies having bBacterial cells in the CRYEMA medium are stained with carbol fuschin and visualized under a compound microscope.
   - *Glucose Peptone Agar Test (GPA Test)*: Rhizobial colonies are streaked on YEMA medium and a master plate is made. Colonies in the

master plate are transferred to GPA medium in a petri dish by replica plating. Rhizobium cannot grow in the medium, but Agrobacteria, if present in the colonies, grow into colonies. Those colonies in the master plate failed to grow in the replica plate are pure rhizobial colonies. They are picked up and grown in YEMA medium. This test is the confirmative test to test the purity of Rhizobial colonies.

- *Nodulation Test*: The same species of legume from which Rhizobia were isolated, is grown in sterile soil in sterile jars. Nutrient solution that lacks nitrogen source, is supplied through a hole at the base of the jars by inserting a cotton thread immersed in the solution. Each pure Rhizobial culture is then inoculated near the root of legume growing in a jar. After 3 or 4 weeks, the plants are carefully uprooted from the jars and visualized for root nodules. Presence of more root nodules indicates that it is the right strain of Rhizobium for that legume.

3. *Establishing the Starter Culture:* Pure rhizobial colony is transferred to a flask containing YEMA medium. The flask is kept on a rotary shaker system in a constant temperature room at 28±2°C. Pure culture of rhizobium appears within a week. It is known as starter culture or mother culture. A starter culture is frequently sub-cultured for maintaining it for a long time.
4. *Mass culture of Rhizobium:* Rhizobium is mass cultured in large bioreactors (fermenter) to prepare inoculant. YEM medium * or sucrose-mannitol medium is used for this purpose.

A suitable medium is formulated and filled into the bioreactor after proper sterilization. One litre of starter culture for 100 litres of medium is inoculated into the bioreactor.

The temperature inside the reactor is maintained at 28± 2°C. Sterile air is continuously supplied to the broth with the help of a proper device.

Cell counting is done a regular intervals to assess the growth rate of Rhizobium. Having reached 10^8-10^9 cells/ml, the broth is harvested to use as inoculant. Further maintenance of the broth in the bioreactor may lead to death of rhizobial cells due to the deficiency of enough nutrients.

TECHNOLOGY ASPECTS OF BIO-FERTILIZERS

What are Bio-fertilizers

- Bio-fertilizers, in strict sense, are not fertilizers which directly give nutrition to crop plants. These are cultures of micro organisms like bacteria, fungi, packed in a carrier material. Thus, the critical input in Biofertilizers is the micro organisms. They help the plants indirectly through better Nitrogen (N) fixation or improving the nutrient availability in the soil.

Mode of Action

- The mode of action depends on the species of the organism. Some agents like Rhizobium cultures enhance N fixation in legumes by imparting effective modulation as they are symbiotic bacteria living in association with leguminous plants.

There are free living bacteria like Azotobacter when applied to soil, enhance the N availability. There are certain other organisms which act on the soil minerals and dissolve the native nutrients like P which is otherwise not readily soluble.

The most commonly produced and marketed biofertilizers are:

- Rhizobium - 3 to 4 strains
- Azotobacter
- Azospirillum
- Phosphate Solubulizing Bacteria - 2 species.
- Blue Green Algae and Azolla - on-farm level.
- VAM - in a limited way.

Critical Factors Responsible for Effectiveness

The critical factors which are responsible for the effectiveness of a particular bio-fertilizer are as follows:

- Suitability of the species to the target crop
- Suitability of the strain: There are specific strains of rhyzobium for different leguminous species like Cowpea, Redgram, Soybean, Alfalfa etc. Biofertilizer of specific culture should be used for specific crop.
- Identification of strains as suited to the agro-eco system, particularly the soil PH and moisture conditions. Through research, specific strains as suited to a particular soil and environmental conditions are usually identified and pure mother cultures are maintained in research labs for supply to the commercial manufacturers, *e.g.* germ plasm of Rhizobium cultures is maintained at IARI, New Delhi.
- The aseptic conditions of manufacturing, the cell count of living organism present in the carrier material, purity and level of contamination.

- The conditions of carrier material in which the culture is packed and the quality of the packing material, which determine the shelf life.
- The conditions in which the packed materials are stored, distributed and kept with the farmers before it is applied.
- Soil conditions particularly PH, organic matter content and moisture level; and agronomic practices.

Level of Benefits

The benefits usually obtained will not be as visible as that of chemical fertilizers except in some critical conditions. Biofertilizers can add nitrogen from 20 kg/ha to 200 kg/ha depending upon the optimum conditions. Pastures and forages respond more than grain crops. The yield increases usually range around 10-35%. However, in the vast areas of low input agriculture and in the context of imparting sustainability to crop production at reduced chemical pollution, this product will be of much use. Rhizobium culture treatment becomes essential when new types of legumes like Soybeans are introduced in new areas. The range of benefits usually seen for different organisms is also given in Annexure 1.

Other Benefits

Application of biofertilizers should not be viewed from the only angle of nutrient supply to the crops. They add life to the soil rendered sterile by the excess use of chemicals, etc. Some of them possess growth promoting substances and also reduce the incidence of certain diseases. These inputs are crucial if some one would like to take up organic farming.

The Outlines of Commercial Manufacture of Biofertilizers

The manufacturing process in short involves:

- Selection of suitable strain of the organism for which market demand is identified.
- Mass multiplication.
- Mixing of the culture with carrier material and packing.

The steps involved are as follows:

Culture Selection and Maintenance

The pure mother cultures of various strains are being maintained in Agricultural Universities, IARI, some ICAR institutions, Regional biofertilizer labs of MOA, etc. There are international sources of supply also like NifTAL, IRRI etc. The mother culture in test tubes of desired strain can be purchased from the identified sources. They have to be further sub-cultured and maintained purely for mass production by adopting standard techniques under the supervision of trained microbiologist.

Culture Augmentation

In the next stage the culture has to be mass multiplied in two levels namely:

1. At primary level using shakers in flasks and
2. Secondary stage multiplication in fermenters.

The important factor in this is the preparation of growing medium in which the culture is mass multiplied. There are standard media on which information is available from published sources like Norris and date, Fred*et al.*, ISI approved etc. in case of Rhizobium. Similarly composition for growth media are available for other cultures. After the media is formulated and sterilized in fermenter, it is inoculated using the shorter cultures multiplied in the flasks at definite ratios usually 5%. The bacteria growing medium is called broth and it is continuously aerated by passing sterile air from compressors. After about 3-4 days fermentation period, the broth will be ready for packing in a carrier material. At various stages the quality is tested by drawing samples.

Carrier Sterilization

While the broth is getting ready in the fermenter the carrier material, which is usually the carbon source for the cultures to survive, is sterilized in autoclaves and kept ready for mixing the broth. Peat imported from countries like U.S., Australia is reported to be the best source of carrier material. However, as it is costly lignite is used extensively in India. The carrier is either sterilized in bulk or it is packed and then the packets are sterilized.

Mixing and Packing

There are 2-3 alternatives depends upon the sophistication and automation of the unit:

1. Under non sterile system, the broth is harvested from the fermenter into sterilized carrier - the mixing is done manually under aseptic condition and packed in polythene bags of desired quantity.
2. In a slightly upgraded method, the broth and sterilized carrier are mixed mechanically in a blender and the material is packed using semiautomatic packing and sealing machine. In a slightly modified method some units are packing by delivering desired quantities of carrier and broth simultaneously from separate pipe conveyance system in to the polythene bags.

3. Under a completely sterile system the carrier is taken in autoclavable polypropylene bags and pre sealed - into which the broth from fermenter is directly injected with the help of dispenser. The injection hole is immediately sealed. The packets are kept in incubation room for about a week before transferring to store room.

Sterile system of packing using auto syringe and dispenser is recommended to be the best method and all new units should follow and adopt this system.

Equipment Needed

The main equipment needed for manufacture and lab are listed below. They are available through scientific and lab equipment suppliers.

- *List of equipment*
- Name Approximate Purpose
- Quantity needed
- Boiler/steam generator 1 big or 2 small To generate steam for sterilization
- Autoclaves - Horizontal 1 For carrier sterilization Vertical 2-4 For smaller quantities and small containers.
- Rotary shakers (2 tier) 2 Culture growth
- Fermenters 2 for culture fermentation (500 lit capacity)
- Laminar air flow -work station of 6' size 2 for inoculation purposes
- BOD incubator 1 for culture growth sterilization
- Hot air oven 2 for dry glassware
- Air conditioner 3-4
- Refrigerator 2
- Microscope 1
- Balances 2-3
- Dispensers or Semi automatic mixing 2 sets with automatic injection systems
- Sealing machine 1
- Lab equipments: For quality control and microbial works
  - pH meter 1
  - Colony counter 1
  - Microscope 1
  - Fridge 1
- Glassware As needed. Conical flasks are the major requirement

- Distiller water unit or 1 set Demineralization unit
- Office furniture As needed

Layout of the Production Unit

The biofertilizer plant should be housed in a suitable building complex. The main production unit should have separate channels for bacteriological work, carrier making and mixing and customer and visitor/marketing way.

In addition there should be rooms with separate entrance for utilities like power, steam generator and stores. Appropriate design can be adopted in consultation with scientists/engineers.

Raw Material

The chief raw materials needed for the production of biofertilizers are as follows:

- Mother cultures
- *Carrier material*: Lignite or bentonite or peat of desired quality in powder form (70-100 mesh)
- Polythene bags, HDPE bags, cardboard cortans
- *Growth materials*: Include Manital, sucrose and chemical nutrients.

Others

In addition to the equipment given in para 6.6, the unit may also require 2 transport vehicles. (1 LCV and 1 Jeep)

Quality Control

Though there are BSI standards for two species *viz.* Rhizobium (IS:8268-1976 and Azotobacter (IS:9138-1979), there is no systematic quality certification system and monitoring mechanism. It is entirely an internal arrangement and voluntary system as of now. As the products being living microorganisms, the quality check up, certification batch-wise even if it is internal is highly essential. Each unit should have lab infrastructure and plans/arrangements for the same.

Each unit, therefore should have the following facilities:

- Adequate microbiological lab and qualified microbiologist.
- Sampling and testing at various stages of production, including the quality of raw materials.
- Specify on the packets all the contents and cell counts. The source of mother culture and the strain name should also be mentioned.
- The unit should fix their quality certificate and batch number, pack the products in proper packing material.

- Store the products in cooler places till they are sold to farmers.
- Ensure to have aseptic conditions, cleanliness and contamination free production lines and housing.
- Preferably use automatic and closed systems.

As per BIS specifications, certain tests are required to be conducted, like no of cells, colony character, reaction etc. Cell number at the time of manufacture should not be less than 10^8 and 10^7 per gram of carrier material, respectively for Rhizobium and Azotobacter. Similarly, the number of cell count and permissible contamination at expiry dates are also specified. As certification arrangements are not in place at present, legislation for quality monitoring and accredited labs for testing may be needed in future to ensure proper quality and promote this products.

Limitations and Constraints

The major limiting factors include:

- Narrow genetic base of mother cultures and lack of efficient and virulent strains suitable to various agro-environments.
- Unsatisfactory carrier material with uniform and consistent good quality comparable to imported peat material.
- Contamination in broth mixing and packing stages, not using completely closed system of production.
- Unsatisfactory packing material which reduces shelf life.
- Unsatisfactory storing conditions, particularly during the distribution period. Exposure to high temperatures and sunlight destroy the microbial culture. They should be preferably kept in cold storage conditions.
- Not employing properly trained microbiologist.
- Lack of quality controls and certification procedures.

TYPES OF ORGANIC FERTILIZERS

Animal Manure

Animal manures vary widely in nutrient content and nutrient availability, depend-ing on the type of animal that produced the manure, and the age and handling of the ma-nure. Mixing with bedding dilutes manure. Exposure to rain leaches nutrients. Composting under cover retains more nutrients, but reduces nutrient availability. Table compares typical nutrient values of different manures. The nutrient content of manure could differ substantially from the amounts in this table, depending on how the manure was stored and handled. Applying farmyard manures. It doesn't take much of a nutrientrich manure, such as poultry manure, to fertilize a garden. A 5-gallon bucket of poultry manure contains enough nutrients to fertilize 100 to 150 square feet of garden. If more is applied, you risk over-fertilizing, harming crops, and leaching nitrogen into groundwater.

Dilute manures, such as separated dairy solids and horse manure with bedding, contain far fewer available nutrients and can be applied in larger amounts. You can use as much as an inch of these materials in the garden every year. Use these manures mainly as a source of organic matter.

Experiment with the amount you apply and observe the performanceExperiment with the amount you apply and observe the performance of your crops to fine-tune your application rate. It is better to be conservative and add more manure if the crops appear deficient.

Table. Typical nautrient content of uncomposted animal manures at the time of application.

Type	N	P_2O_5[2]	K_2O
	1b per cu yard as-is[3]		
Broiler with litter	33	29	30
Laying hen	26	40	33
Sheep	13	6	24
Rabbit	11	7	10
Beef	8	4	12
Dry stack dairy	6	3	13
Separated dairy silids			
Horse	6	4	11

1. Divide these number by 40 to estimate the nutrients in a 5-gallon bucket of fresh manure.
2. Phosphorus and potassium in this table are shown in units of izer lables. To convert from P to P_2O_5, multiply P by 2.3. To convert from K to K_2O, multiply K by 1.2.
3. As-is is typical for manure stored under cover.

Commercial manure composts. Composted chicken and steer manures with known nutrient contents are commercially available as bagged products. Table shows typical analyses of these materials.

Nutrient levels and availability in steer manure compost are so low that you should use steer compost as a source of organic matter only. Recommended application rates on packages of these products are good guidelines for their use.

Table. Nutrient content of commercial manure composts.

Animal	N	P_2O_5	K_2O
		%	
Chicken	1-3	0.5	1-2
Steer	0.5	1-2	0.5

Using manure safely. Fresh manure sometimes contains pathogens that can cause diseases in humans. These pathogens are not taken up into plant tissue, but they can adhere to soil on plant roots, or on the leaves or fruit of low-growing crops.

Cooking destroys pathogens, but raw food carries a risk of pathogen exposure. Washing and peeling raw produce removes most pathogens, but some may remain.

The risk from pathogens is greatest for root crops or leaf crops where the edible part touches the soil. The risk is negligible for crops such as sweet corn, which do not come in contact with the soil, or for any crop that is thoroughly cooked.

Consider raw manure to be a potential source of pathogens, and avoid using fresh manure where you grow high-risk crops. Bacterial pathogens die off naturally during composting, extended storage, or after field application. Complete die-off of bacterial pathogens occurs in days to months depending on the pathogen and environmental conditions.

Keep dog, cat, and pig manure out of your home compost and garden. Some of the parasites found in these manures are very persistent, and may survive in both compost and soil.

The best time to apply manure is in the spring before planting. You also can apply manure in the fall, but some of the nutrients may be lost over the winter. Environmental risks of leaching and run off also increase. If you do apply manure in the fall, apply it early, and plant a cover crop to help capture nutrients and prevent run off.

Don't over apply fresh manure. Because fresh manure contains available nutrients as well as slow-release nutrients, over application can harm your crops.

A safe way to use fresh manure is to mix it into your compost pile and then use the finished compost in your garden. The micro-organisms breaking down the other materi-als in your compost will absorb the available nutrients, reducing the risk of nutrient over-application and leaching.

Because it is hard to maintain the high temperatures needed to kill pathogens quickly in a backyard compost pile, allow your manure compost to age for six months or more before using it in your garden. An alternative is to buy commercially composted manure.

Biosolids

Biosolids are a by-product of municipal wastewater treatment. Federal standards for organic farming do not include biosolids as an organic fertilizer.

However, biosolids do have two important characteristics of organic fertilizers:

1. Their nutrients are released slowly from the organic form by natural processes in the soil, and
2. They are a product of the waste stream that can benefit crop growth.

Most of the biosolids produced in Washington are used to fertilize agricultural and forest crops. The material used on farms is rich in nutrients and acts similarly to poultry manure.

Some communities in Washington produce a special class of biosolids and market them to gardeners. These are called Class A biosolids: they have been treated using heat or composting to reduce pathogens to background levels, making them safe for all garden uses.

Three types of Class A biosolids are available in Washington: composts, blends, and heat-dried pellets. Biosolids composts are made from biosolids and yard debris or woody materials, and can be used similarly to other composts.

Biosolids blends are formulated for different uses, including turf topdressing, mulches, and soil amendments. Heat-dried pellets are rich in nutrients and are used similarly to commercial organic fertilizers.

Commercial Organic Fertilizers

Many organic by-products and some unprocessed minerals are sold as commercial organic fertilizers. Table shows approximate nutrient content of some of these materials. Total nutrient contents; because most are slow-release fertilizers, not all of the nutrients will be available the same year they are applied. The table shows that each fertilizer contains one main nutrient. The other nutrients are present in smaller amounts. Several companies produce bal-anced organic fertilizers, blended into a single product that provides all of the primary nutrients.

Commercial organic fertilizers tend to be more expensive per pound of nutrients than either synthetic fertilizers or manures. Sometimes the difference in price can be substantial. Choosing organic fertilizers involves tradeoffs in cost and convenience.

Farmyard manure is usually inexpensive or free, but is less convenient than packaged, commer-cial materials. If you or your neighbours have livestock it makes both environmental and economic sense to recycle the manure pro-duced by the livestock.

Packaged organic fertilizers can be expensive, but gardeners may choose them where convenience or quick availability of nutrients is important, or for small gardens where little fertilizer is needed. The cost per pound of nutrients in organic fertilizers varies widely, depending on the type of material, the concentration of nutrients, and the size of the package. Compare costs and nutrient availabil-ity when shopping for organic fertilizers.

Table. Total nitrogen, phosphate, and potash content typical of some organic fertilizer.

Meterial	N	P_2O_5	K_2O %
Cottonoseed meal	6-7	2	1
Blood meal[1]	12-15	1	1
Alfalfa	2	0.5	2
Bat guano[1]	10	3	1
Fish meal[1]	10	4	0
Fish emulsions[1]	3-5	1	1
Bone meal	1-4	12-24	0
Rock phosphate[2]	0	25-30 (only 2-3% available)	0
Greensand	0	0	3-7
Kelp meal	1	0.1	2-5

- These materials contain a substantial amount of quickly available nitrogen that plants can use early in the season.
- Very low availability. Useful only in acid soils.

Lime, Gypsum, and Wood Ashes

Lime, gypsum, and wood ashes are all mineral materials used as organic soil amend-ments. Lime and gypsum are commercial products, while wood ashes are produced from home woodstoves and fireplaces.

Lime and pH. Lime is ground limestone, a rock containing calcium carbonate. Lime raises the pH of acid soils, and supplies the essential nutrient, calcium.

Dolomitic lime contains magnesium as well as calcium, and is a good choice for organic gardeners in western Washington, where garden soils often are deficient in magnesium. In strongly acid soils, many nutrients are less available to plants, and some toxic metals are more available to plants.

This can be corrected by the proper application of lime. Soils are naturally acidic in humid areas—most of western Washington—and are neutral to alkaline in drier areas—most of eastern Washington. Fertilizers tend to increase soil acidity over time; some eastern Washington topsoils have become acidic from fertilization.

The best way to determine if your soil needs lime is to have your soil tested. In the absence of a soil test, western Washington gardeners can add about 50 pounds of dolomitic lime per 1000 square feet of garden per year. Do not lime areas where you grow acid-loving plants, because they are adapted to acid soils. Lime is a slow-release material. Fall application will benefit a spring crop.

Gypsum is not a substi-tute for lime. It provides calcium and sulfur to soils, but has little effect on soil pH. Gypsum has been promoted as a soil amendment to improve soil structure.

In the vast majority of cases it will not work. Gypsum improves structure only when the problem results from excess sodium in the soil, a rare condition in Washington. Use organic amendments to improve soil structure, as described earlier.

Wood ashes are a readily available source of potassium, calcium, and magnesium. They also act like lime, raising the pH of the soil. High rates of wood ashes may cause short-term salt injury, so keep applications to less than 15-25 pounds per 1000 square feet of garden. We do not recommend using wood ashes in alkaline soils.

VARIETIES FOR ORGANIC FARMING

Widespread use of improved rice varieties since the 1960s has reduced food prices, for the poor and prevented millions of cases of childhood malnutrition. Without the development of the high yielding varieties, prices for developing country consumers would likely be as much as 40 per cent higher than they are today.

The new varieties have also reduced costly food imports by almost 8 per cent and have eliminated the need to convert millions of hectares of forestland to agricultural uses as would have otherwise been required had yields remained at 1960 levels.

The availability of high yielding varieties has prompted many of the world's poorest countries to invest in plant breeding programmes and produce varieties suited to local environments and markets. The new plant types have also prompted massive government investments in agricultural infrastructure such as irrigation and fertilizer delivery systems.

In spite of the advantages of high yielding varieties, it should not be forgotten that the adoption of high yielding has led to the substitution of a large quantity of species for only a few and uniform varieties from a genetic point of view, which has caused a significant reduction in the genetic inheritance of cultivated species.

Many agricultural species, varieties and breeds which have played an important role in the human diet and traditional cultures have practically disappeared over the last century.

In the last decade, the adoption of organic agriculture has indirectly established a rescue process of species, varieties and breeds threatened by under-use or extinction.

Stronger collaboration has been evident among movements aiming to defend biodiversity and the organic agriculture movement. This is especially the case now that there is interest in traditional, speciality and organic products.

For the rescue of varieties threatened by extinction, the development of a market is fundamental and it is here that organic agriculture plays an important role as the price premium gives an additional value to the product.

The restoration and enhancement of under-utilized species and varieties has been motivated by a food demand concerned with health and culinary traditions. Organic agriculture has allowed the maintenance and improvement of species and varieties that otherwise would suffer strong genetic erosion or extinction.

Organic Agriculture

Organic systems avoid the use of synthetic fertilizers, pesticides, and growth regulators. Instead they rely on crop rotations, crop residues, animal manures, legumes, green manures, off-farm wastes, mechanical cultivation, mineral-bearing rocks, and biological pest control to maintain soil health, supply plant nutrients, and minimize insects, weeds, and other pests.

Selection of Rice Varieties for Organic Farming

In general, rice planting dates, seeding rates, preferred varieties, and harvesting methods vary among regions, but they are largely the same for conventional and organic systems While choosing varieties for organic farming, the special considerations relevant to *organic* rice production has to be duly recognized. Weed control, management of soil fertility and minimizing pests and diseases are the principal challenges associated with organic rice production and according to the needs and requirements in each, varieties have to be chosen.

Weed Control

Weeds are a problem in both wetland rice as well as up lands, but it is more hazardous under up land conditions. Direct seeding of rice is receiving much attention in recent times since it reducesproduction costs. As cultural practices for rice shift from transplanted to direct seeded rice, weed problems will increase because rice and weeds can emerge together.

Many weeds of irrigated ecosystems like Barnyard grass *(Echinocloa crusgalli)* can become more important /harmful in direct seeded rice since they are adapted for better growth under dry than wet conditions. While choosing varieties for upland rice, care should be taken to select varieties with initial vigour and increased growth rate so that they can compete with weeds and come up well. Short duration varieties are preferred for direct seeding in uplands, since they have higher growth rate compared to medium or long duration varieties.

Girija *et al.* (2002) while evaluating upland varieties for their weed competitiveness at Regional Agricultural Research Station, Pattambi, noticed distinct difference in the weed suppression ability of the cultivars tested.

Based on this, they classified the rice varieties into three groups viz.

1. Varieties, which can completely smother the weeds,
2. Varieties that show moderate tolerance and
3. Varieties that are smothered by competing weeds.

Improved traditional varieties like PTB 28, PTB 29 and PTB 30 and land races like Karanellu belong to the first group. The improved upland varieties could grow fast and produce large no. of tillers, there by preventing the growth of weeds in between. Land races like Karanellu had long and droopy leaves, which prevented light interception by weeds there by reducing their growth. Research is underway to transfer the weed competitiveness of these varieties to high yielding varieties.

In transplanted rice in lowlands, crop rotations, land leveling, seedbed preparation, water management, and rotary hoeing are the primary weed-control practices followed to take care of the weeds. Here also, varieties having natural weed fighting capacity could be chosen to reduce the hazards from high weed growth. Use of allelopathic rice varieties having activity against a broad spectrum of target weeds could be a desirable choice.

Cultivars that show allelopathic potential against important rice weeds have been identified in many countries *viz.*, United States, Japan, Egypt and the Philippines. In the United States, evaluated more than 10000 accessions of rice for allelopathic effects against aquatic weeds like duck salad and red stem and could identify around 190 promising accessions. This included rice varieties from many countries including India. Allelopathic cultivars that strongly inhibit root elongation of barnyard grass but weakly affect the shoot have been identified by Olofsdotter. In Egypt, identified Indian rice varieties that expressed allelopathic effects on Echinocloa cruss-galli and Cyperus difformis. L. at 3-4-leaf stage by inhibiting root development and emergence of the first or second leaf of both weeds. Kim and Shin, 1998 identified promising rice germplasm for allelopathic activity in Korea and new cultivars are being generated using them.

Soil Fertility

Maintaining soil fertility in organic cropping typically involves some combination of crop rotation with deep-rooted legume crops or green manure/ cover crops, and applying rock minerals, animal manures, composts, and other approved organic amendments. Green manures have proven their positive influence in enhancing rice yields. Leguminous green-manure crops can supply 30 to 50 per cent of the nitrogen needs of high yielding rice varieties. The availability of green-manure nitrogen depends on the quantity, quality, and type of green-manure crop; the time and method of application; soil fertility; and cropping method.

It has been observed that green manuring with Sunnhemp or Cowpea grown during summer months and incorporated in soil before transplanting rice could save about 60 kg N/ha for rice crop. Recent studies on economizing chemical fertilizers through organic manures in selected cropping systems under All India Co-ordinated Agronomic Research Project have shown that 25-50% N requirement of rice during Kharif season could be met through organic sources without any adverse effect on rice productivity. Yield response of rice varieties to organics indicated varietal differences to organic farming. Some varieties responded more to organics than to chemical fertilizers. Similarly, certain rice varieties responded to chemical fertilizers than to organics.

Emission of trace gases especially Methane from flooded rice fields has become a serious concern world over. Methane emission from lowland rice has been identified as resulting from organic matter fermentation and factors like incorporation of crop residues like rice straw can enhance the process. Varieties with low methane emission potential could reduce 'hazards resulting from high methane emission and could be more rewarding under organic farming.

Insects and Diseases

Rice, both rain fed and irrigated is infested by a variety of pests and diseases. Intensive cropping of modern varieties under high inputs provided a favourable environment for many pests as a limited number of modern varieties began to be widely cultivated, the populations and economic importance of their associated pests increased and new pathogen strains and insect biotypes developed. The major insect pe.sts *viz.*, Plant hoppers gall midge, stem borer, leaf rollers, rice bug, along with minor pests like case worm, thrips, ear cutting caterpillar, whorl maggot etc., cause serious damage to the rice crop in the tropics. Similarly, fungal diseases (leaf blast, sheath blight, brown spot, sheath rot, bacterial diseases (bacterial blight) and viral diseases (Tungro Virus, Grassy Stunt Virus) etc., can also devastate the crop. Excessive nitrogen levels are rarely a problem in organic production, since use of synthetic fertilizers is avoided, but timely control measures are a must to keep the damage under economic threshold levels. Timely planting, variety selection, and cultural practices to suppress weeds and encourage dense stands of rice will help control most of the biotic stresses mentioned above.

Use of resistant varieties can prove to be consistently and significantly more productive under high pest or disease pressure. Varieties with moderate resistance are a better choice since they allow pest populations to be maintained at levels that do not result in significant damage. Varieties with strong resistance reduce the populations of natural enemies that feed on the pest and could ultimately lead to resurgence of the pest. List of high yielding varieties combining resistance to various biotic stresses released by the Central Seed Subcommittee on Crop Standards, Notification and Release of Varieties. The different states have also released a number of varieties resistant to various biotic stresses.

Selecting varieties possessing multiple resistances could make the programme more effective. Again, the economic loss associated with each stress should be assessed before selecting a variety and high priority should be assigned to varieties resistant to a particular stress for which biological control measures are not available, and low priority could be given for varieties resistant to such stresses where other low cost control methods are available. For eg. Use of varieties resistant to Bph and gall midge and adopting biological control measures for stem borer and leaf roller could take care of all the above pests. Similarly, selecting a blast resistant variety along with biological control measure for sheath blight could be a more effective method for controlling the two diseases rather than going for varieties resistant to both.

Bibliography

Badiru, A.: *Handbook of Industrial Systems,* CRC Press, Delhi, 2005.

Bahar A. Siddiqui and Samiullah Khan: *Plant Breeding Advances and in vitro Culture,* CBS, Delhi, 1997.

Banki, L.: *Bioassay of Pesticides in the Laboratory,* Akademiai Kiado, Budapest, 1978.

Barrington, E. J. W.: *Biochemistry of Primitive Deuterostomians,* London, Academic Press, 1974.

Benvenuto, E.: *An Introduction to the History of Structural Mechanics,* Springer-Verlag, New York, 1991.

Brodwin, Paul : *Biotechnology and Culture: Bodies, Anxieties, Ethics,* Bloomington: Indiana University Press, 2001.

Collins, Steven : *The Race to Commercialize Biotechnology: Molecules, Market and the State in Japan and the US.,* New York: Routledge, 2004.

Crozet, Francois: *The First Industrialists: The Problems of Origins,* Cambridge University Press, New York, 1985.

Davenport, Thomas: *Process Innovation: Reengineering Work through Information Technology,* Harvard Business School Press, Boston, 1993.

Doijode S. D.: *Seed Germination in Fruits,* New Delhi, Malhotra Publishers, 1993.

Drucker, Peter: *Work and Tools,* Technology and Culture, New York, 1972.

Elliott, B N : *Biochemistry and Molecular Biology,* Oxford University Press, Delhi, 2005.

Flond, R.: *The British Machine-tool Industry: 1850-1914,* Cambridge University Press, New York, 1976.

Francis Durso, Patricia DeLucia: "Engineering Psychology", *The Corsini Encyclopedia of Psychology,* John Wiley and Sons, NY, 2010.

Fransman M, Junne G, Roobeek A: *The Biotechnology Revolution?,* Oxford, Blackwell, 1995.

Fumento, Michael: *Bioevolution: How Biotechnology is Changing Our World,* San Francisco, Encounter Books, 2003.

Ganguly, Smriti : *Biochemistry of Biomolecules,* Pearl Books, Delhi, 2007.

Ghulam Hassan : *Soil Microbiology and Biochemistry,* New India Publishing Agency, Delhi, 2010.

Gille, B.: *Engineers of the Renaissance,* MIT Press, Cambridge, 1966.

Goodsell, David S.: *Bionanotechnology: Lessons From Nature,* Hoboken, Wiley-Liss, 2004.

Graf, Alfred Byrd: *Advances in Plant Physiology,* Rajat Pub, Delhi, 2008.

Hammer, M. and Champy, J. A.: *Reengineering the Corporation: A Manifesto for Business Revolution,* Harper Business Books, New York, 1993.

Hardy B.: *Biology and Agronomy of Forage Arachis*, Cali, International Centre for Tropical Agriculture, 1994.

Hill, D.: *A History of Engineering in Classical and Medieval Times*, Open Court, La Salle, IL, 1984.

Jeffers P.: *Evaluation of Four Onion Varieties in Montserrat*, Plymouth, CARDI, 1992.

Jones, R. M.: *Plant Resources of South-East Asia*, Wageningen, Pudoc Scientific Publishers, 1992.

Khanna, V K : *Objective Genetics, Biotechnology, Biochemistry and Forestry*, I.K. International Publishing House, Delhi, 2008.

Kovach, S.P.: *Injection of Fertilizers into Drip Irrigation Systems for Vegetables*, Univ. Florida, Gainesville, USA, 1984.

Kunelius T.: *Annual Ryegrasses in Atlantic Canada*, Ottawa, Agriculture Canada, 1991.

Macself, A.J.: *Soils and Fertilizers*, Satish Serial Pub, Delhi, 2005.

Madan Lal Bagdi: *Physiology, Biochemistry and Biotechnology*, Manglam Pub, Delhi, 2007.

Madulid Domingo A.: *A Pictorial Cyclopedia of Philippine Ornamental Plants*, Philippines, Makati Metro Manila, 1995.

Mahindru, S. N.: *Food Safety and Pesticides*, APH, Delhi, 2009.

Meena Francis: *Biotech's Dictionary of Biochemistry*, Biotech Books, Delhi, 2007.

Mondal, R.C.: *Appropri-ate Technology*, Intermediate Technology Publications, London, 1974.

Monteith, J.L. *Applications of Soil Physics*, Academic Press, Florida, USA, 1980.

Musson, A. E. and Robinson, E.: *Science and Technology in the Industrial Revolution*, University of Toronto Press, Toronto, 1969.

Narasimha Rao, P.: *Irrigation Development: Issues and Challenges*, Discovery, 2007.

Old, R.W. : *Principles of Gene Manipulation*, London, Blackwell Scientific Publications, 1989.

Oldham P.: *Cost of Production of Major Tree Crops in Dominica*, Roseau, Ministry of Agriculture, 1991.

Orson, W. Israelsen: *Irrigation: Principles and Practices*, Axis Books, 2010.

Parry M.L.: *Climatic Change, Agriculture and Settlements*, Dawson Folkestone UK, 1978.

Paul M. Althouse: *Introduction to Agricultural Biochemistry*, Biotech Books, Delhi, 2005.

Pemberton, R. W.: *Predictable Risk to Native Plants in Weed Biological Control*, Oecologia, 2000.

Petr, Tomi: *Fisheries in Irrigation Systems of Arid Asia*, Daya, 2007.

Ragone D.: *Breadfruit: Artocarpus Altilis (Parkinson) Fosberg*, Rome, International Plant Genetic Resources Institute, 1997.

Rifkin, Jeremy: *The Biotech Century*, New York, Penguin Putnam, 1998

Sharma, Pradeep : *Biochemistry and Organisation of Cells*, RBSA Pub, Delhi, 2006.

Turner, W.: *Introduction to Industrial and Systems Engineering*, Prentice Hall, NY, 1992.

Urton, Gary: *The Social Life of Numbers*, Austin, University of Texas Press, 1997.

Vanangamudi, K.: *Principles and Methods of Plant Breeding*, International Book, Delhi, 2005.

Wickens, C. and Hollands, J.: *Industrial and Agriculture Performance*, Prentice Hall, NY, 1996.

Index

❑❑❑